The 6G Manifesto

William Webb

ISBN-13 9798338481936

© William Webb, 2024

First published October 2024

Manifesto

The British and American English word "manifesto" derives from the Latin terms "manifestus" and "manifestum," both of which mean "obvious." Over time, the Latin morphed to "manifesto" ("to make public") before changing again to a new word in Italian with the same meaning: "manifestare." Beginning in the seventeenth century, English speakers borrowed the Italian word and returned it to the Latin "manifesto," using it to mean a written document that elucidates beliefs and calls for change.[1]

---

[1] https://www.masterclass.com/articles/what-is-a-manifesto

The 6G Manifesto

Table of Contents

Structure of this book 3
1   Do we still need generations? 5
   1.1   Why we might need generations 5
   1.2   Within-generation updates 9
   1.3   What makes a new generation? 9
   1.4   Why we will get 6G regardless 10
2   Lessons from 5G 13
   2.1   A generation like no other 13
   2.2   Solution in search of a problem 15
   2.3   Poor choice of frequency bands 16
   2.4   Politics and the race to 5G 19
   2.5   Expenses imposed on MNOs 20
   2.6   A hype machine in overdrive 21
   2.7   Is there any hope for 5G? 21
   2.8   Implications for 6G 22
3   The problems that are clearly visible 23
   3.1   Ask and you will be told 23
   3.2   Ensuring users are always connected 25
      3.2.1   Introduction 25
      3.2.2   Urban connectivity 27
      3.2.3   In-building connectivity 29
      3.2.4   Rural connectivity 32
      3.2.5   Travel connectivity 40
      3.2.6   Capacity issues 41
      3.2.7   A role for neutral hosts? 42
   3.3   Reducing costs 44
   3.4   Achieving both simultaneously 46
4   Current 6G visions 47
   4.1   Introduction 47
   4.2   Manufacturers 47
      4.2.1   Ericsson 47
      4.2.2   Huawei 49
      4.2.3   Nokia 51
      4.2.4   Samsung 53

	4.2.5	The focus on the cyber world	55
	4.2.6	Manufacturers summary	64
4.3		Operators and operator alliances	65
	4.3.1	NGMN	65
	4.3.2	Operator remarks	71
	4.3.3	Operators summary	72
4.4		Research bodies	72
	4.4.1	University of Surrey / 6GIC	72
	4.4.2	6G Flagship	73
	4.4.3	Academia summary	75
4.5		Alliances and similar	75
	4.5.1	NextG Alliance	75
	4.5.2	6G Industry Association (6GIA)	76
	4.5.3	ITU	79
	4.5.4	Summary for associations and similar	84
4.6		Overall summary of visions	84
5		Assessing what the visions call for	85
5.1		The visions are not clear on what is needed	85
5.2		Calls for more spectrum abound	86
5.3		AI-native is trendy but vague	88
5.4		Sensing is new but ill-conceived	89
5.5		It adds up to more expense than 5G for less inclusivity	91
6		A better 6G vision	93
6.1		Introduction	93
6.2		What is not needed	93
6.3		What 6G needs to deliver	94
6.4		Achieving multi-network operation	94
6.5		The multi-network coordinator	97
6.6		Protocols (by David Lake)	102
6.7		Who owns the multi network coordinator?	106
6.8		Economics	109
6.9		Security	110
6.10		Lowering energy usage	111
6.11		Legal requirements on operators	112
6.12		The optimal outcome	112
7		How to get to the optimal outcome	114

Table of contents

	7.1	Where we are heading	114
	7.2	Operators versus manufacturers	116
	7.3	Issues of national sovereignty	117
	7.4	Challenges in delivering a multi-network coordinator	121
	7.5	Government support is likely to be needed	122
	7.6	Time for the operators to make a stand	122
8	The book in overview	124
	8.1	The long summary	124
	8.2	The short summary	129
9	The manifesto	131

List of abbreviations

3GPP	Third Generation Partnership Project
5GIC	5th Generation Innovation Centre
6GIA	6G Industry Association
6GIC	6th Generation Innovation Centre
AI	Artificial Intelligence
API	Application Program Interface
AR	Augmented Reality
CDMA	Code Division Multiple Access
CEO	Chief Executive Officer
CMA	Communications and Market Authority (UK)
CPE	Customer Premise Equipment
C-RAN	Cloud-RAN
D2D	Direct to Device (from a satellite)
DASH	Dynamic Adaptive Streaming over HTTP
eMBB	Enhanced Mobile Broad Band
EMF	Electro-Magnetic Field (safety measures)
ENUM	Enumeration
ESG	Environmental, Sustainability, and Governance
FWA	Fixed Wireless Access
GPRS	General Packet Radio Service
GTP	GPRS Tunnelling Protocol
HAPs	High Altitude Platforms
HD	High Definition
IETF	Internet Engineering Task Force
IoT	Internet of Things
IMEI	International Mobile Equipment Identity
IMT	International Mobile Telecommunications (cellular)
ITU	International Telecommunications Union
IP	Internet Protocol
KPI	Key Performance Indicator
LEO	Low Earth Orbiting (satellite network)
MBB	Mobile Broad Band
MIMO	Multiple Input Multiple Output
ML	Machine Learning
mMTC	Massive Machine Type Communications

MNO	Mobile Network Operator
MOCN	Multi-Operator Core Network
MRSS	Multi-RAT Spectrum Sharing
MS	Mobile Service (frequency band)
MSS	Mobile Satellite Service (frequency band)
MVNO	Mobile Virtual Network Operator
MWC	Mobile World Congress
NDN	Named Data Networking
NGMN	Next Generation Mobile Network (Alliance)
NR	New Radio
NSA	Non Stand Alone (5G core)
NTN	Non Terrestrial Network
OFDM	Orthogonal Frequency Division Multiplexing
O-RAN	Open RAN
QAM	Quadrature Amplitude Modulation
QUIC	Quick UDP Internet Connections
R&D	Research and Development
RAN	Radio Access Network
RF	Radio Frequency
RIC	RAN Intelligent Controller
RINA	Recursive Internetworking Architecture
RIS	Reflective Intelligent Surfaces
SA	Stand Alone (5G core)
SCS	Supplementary Coverage from Space
SDO	Standards Development Organisation
SLAM	Simultaneous Localization And Mapping
SNS JU	Smart Networks and Services Joint Undertaking
SRN	Shared Rural Network
TDMA	Time Division Multiple Access
UAM	Urban Air Mobility
URLLC	Ultra-Reliable Low Latency Communications
VoLTE	Voice over LTE
VR	Virtual Reality
WRC	World Radio Conference
XR	Extended Reality (a combination of VR and AR)

Structure of this book

This book sets out a vision as to what 6G should be – a manifesto proposing an alternative view to much of what has been published to date.

The first chapter discusses the concept of mobile generations, looks at what defines a new generation and concludes that, while there are strong arguments that we do not need another generation, the 6G "super tanker" has already left the port and hence the arrival of 6G is inevitable. But if that is so, then at least it should be a generation that benefits users and stakeholders.

The second chapter looks at the lessons that can be learned from 5G. It shows that 5G is, at least to date, a failure in that it has not delivered on its vision, it has not been profitable, and users have not noticed any benefits. It discusses the factors that caused this including the excessive hype, enlisting of politicians and a failure to learn lessons of previous generations. It draws out implications for 6G including that there should be a focus on frequency bands below 3GHz and the need for mobile network operators (MNOs) to have a strong voice in setting direction.

The third chapter looks at the problems with current mobile networks and shows these are a lack of widely available and high-quality coverage, and in some cases high costs. It shows how much improved coverage can be delivered at low cost by utilising other networks including satellite, Wi-Fi and national roaming across other cellular networks.

The fourth chapter looks at the published 6G visions from manufacturers, operators, research bodies and industry alliances. It shows that there are two completely separate camps – the manufacturers and academics arguing for 5G-on-steroids, and the operators pleading for a software only cost-reducing update which improves coverage. The fifth chapter then shows how the manufacturers' visions will lead to an even worse outcome than 5G with very limited 6G availability.

The sixth chapter sets out a better vision for 6G. It calls for always-connected quality coverage, including in rural areas, urban not-spots, indoors, and in trains.

It sets out the need for highly reliable networks but also for lower consumer costs, likely achieved through lower operator capex and opex including reduced energy consumption. It discussed how this vision can best be achieved with a multi network coordinator that sits outside of individual networks and coordinates and provides common services across all of them. It includes a contribution from David Lake on protocols.

The seventh chapter looks at how to bring about this vision and argues that MNOs need to take a strong lead in telling politicians, journalists and the public about a vision that will bring clear benefits to all users, and then they need to pressure manufacturers to ensure that the standards deliver appropriate solutions.

Chapter eight provides a summary of the book while the final chapter summaries the vision in the form of a manifesto.

1 Do we still need generations?

1.1 Why we might need generations

The history of mobile communications is dominated by generations. From 1G through to 5G, new generations have brought new features, enhanced capacities and sometimes new operators. The 10-yearly cadence of generations has set the way that many work - from regulators to manufacturers to academics.

I have started many of my recent books with a short summary of 1G through to 5G so will avoid doing so here, but I will pick up on certain generation changes throughout this chapter. Indeed, this book touches on ideas and issues discussed in detail in recent books including "The 5G Myth" (2016), "Emperor Ofcom's New Clothes" (2024) and "The End of Telecoms History" (2024)[2] and rather than repeat that material, I'll reference those books as needed.

The generational approach, with its regular cadence and significant change across an entire industry, is unusual. Few other areas have anything similar – for example cars, planes, the Internet, computers, chipsets and VR headsets do not have any sort of industry-wide generations. Instead, manufacturers update products when they see fit, perhaps using new model numbering, eg iPhone 12, iPhone13, but this is not done in synchronisation or alignment with other manufacturers. In the case of the Internet there is occasional significant change proposed, such as IPv6 when there is a view that it is needed. In the wireless communications space, Wi-Fi does have something akin to generations, although that may be partly to rival cellular.

There are a few reasons why, at least historically, generations have been helpful:

1. *Harmonised spectrum.* Cellular has needed a steady stream of new spectrum to cope with the huge increases in data volumes that users have demanded. The advent of a new generation has tended to focus regulators and others on achieving international harmonisation and

---

[2] All available on Amazon

implementing an auction process in a timely and somewhat coordinated manner. While it would have been possible to do this without any new generation, and indeed some countries have implemented occasional spectrum awards between generations, having a new generation is a sure-fire way of delivering more spectrum. I will return later to the fact that (1) this is not always useful spectrum and (2) the need for new spectrum may now have diminished or even disappeared.

2. *Focused research.* New generations have sent signals to researchers, both in academia and industry, to look into particular topic areas that are thought of as important. Generations have also more recently resulted in funding from bodies such as the EU tied directly to the new generation. While some of this research may have happened anyway, it is possible that the generational focus has advanced research faster and with greater innovation than would otherwise have been the case. However, if the topic areas called for by those designing the next generation are inappropriate, as I have argued was the case with 5G, then this can result in wasteful research which can preclude work in areas that would have been more useful.

3. *Compatibility.* Standards are essential where two or more devices from different suppliers need to interwork. This is clearly true in cellular where handsets from companies like Apple and Samsung need to work with base stations from companies like Ericsson and Nokia. Compatibility is especially important where it is embedded within hardware, such as the antennas and filters within a handset, since these cannot be changed after manufacture. Conversely, where the compatibility is in software, eg on a laptop computer, then new "drivers" can be downloaded or new internet protocols added dynamically. But this is a weak argument for generations of cellular, since standardisation can occur without a generational approach. The generational approach is only helpful if there is clear need for large-scale update of the standard and that this update needs to be handled across many countries and suppliers at the same time.

4. *Economies of scale.* Somewhat akin to standards, cellular requires economies of scale since devices are extraordinarily complex and the development costs need to be amortised over billions of units for them

to be affordable. Good standards can lead to economies of scale. But poor standards may result in wasted investment or a lack of economies as many choose not to implement them.

What is clear is that, in a somewhat circular argument, generations are very helpful if there is a need for generations. To expand on this, if there is a clear need for improvement over the current generation, and if the focus of the new generation is meeting that need, and if the improvement requires substantial change, then a generational approach fits well with the step-change needed.

Pulling this apart somewhat, a new generation is a good idea when all the following are true:

1. The existing generation does not meet clear needs and cannot be adapted via relatively minor changes to do so.
2. The broad direction to meet those needs (eg via better antennas) is agreed and appropriate allowing focused research.
3. The approach is economically viable for all stakeholders in the value chain.
4. More spectrum is needed to deliver increased capacity (or other purposes).

The table below shows how previous generations have fared against these criteria, starting with 2G (since there was not really a 1G standard).

	Unmet needs	Clear direction	Economically viable	Spectrum needed
2G	Capacity, security, roaming	Digital, TDMA	Yes	Yes (900MHz, 1800MHz, etc)
3G	More voice capacity, then data above 100kbits/s, IP focus	Somewhat, CDMA, IP core	Somewhat, but wrong frequency band, auction fees overpaid	Yes (2.1GHz, 2.6GHz, etc)
4G	Delivering 3G objectives well	OFDM	Yes, lower frequencies, minor network change	Yes (800MHz etc)
5G	No	No	No. Wrong frequencies, no revenue or cost savings	Yes (predominantly 3.5GHz, also 700MHz, 26GHz etc)

Table 1-1 : Criteria for a new generation

Some of these points will be covered in more detail in later chapters, with a whole chapter dedicated to the reasons why 5G has not delivered against its promises. The overall message here is that from 2G to 4G the generational approach was appropriate. But this approach was wrong for 5G and indeed, as will be shown, likely one of the reasons it failed.

The table shows that new spectrum has been needed for all these generations. However, as I set out in "The End of Telecoms History", there is strong evidence that this is no longer the case. Growth in data usage is slowing by around 5% a year from its level of 20% in 2024. If this trend continues, as seems likely, then it will reach zero (ie usage will plateau) well before 6G is deployed. This is highly material to what 6G should be, and fortunate since identifying new spectrum has become progressively harder and is now deeply challenging.

1.2 Within-generation updates

The alternative to generational updates is intra-generational updates. Indeed, these have been a feature of cellular since 2G, with major updates to deliver things like packet data (GPRS) through to today where standalone 5G (SA) was delivered through a new release of the 3GPP standards. There is now a regular release cycle (eg R17, R18) and a new generation is earmarked as one of these releases. Clearly it would be possible to simply continue with intra-generational releases and never have a new generation.

1.3 What makes a new generation?

With regular releases for the existing generations (around every 18 months), there can be a lack of clarity as to what constitutes a new generation.

Historically this has been clear - it has been a completely new air interface. 2G was Time Division Multiple Access (TDMA), 3G was Code Division Multiple Access (CDMA) and 4G was Orthogonal Frequency Division Multiplexing (OFDM). Changing the air interface required substantial modifications throughout the standards and could not be handled by software updates either to devices or networks. These changes to the air interface were made as experience showed existing approaches could not deliver the features needed, and as research found better approaches.

However, this clear distinction broke down with 5G. It continued with the OFDM air interface used in 4G. In a very Orwellian manner, it tried to disguise this by terming its air interface "new radio", when it was anything but. 5G adds additional flexibility to the 4G OFDM interface with broader bandwidths and an increased number of slot sizes, but makes little fundamental change. Most current deployments of 5G rely on non-standalone (NSA) operation where 4G delivers all the device control. 5G could have been an intra-generational enhancement to 4G rather than a new generation.

Despite much research into new air interfaces over the decades, and many alternative proposals, there is nothing that looks likely to displace OFDM as the air interface of choice. Hence, the "new air interface distinction" will not apply to a new generation.

When 5G SA is deployed there will be a very material change in the core network compared to 4G. But since this does not change much on the device side then its impact is internal to mobile operators and there is little need to label this a new generation. It also does not need new spectrum, much in the way of research, or many of the other benefits that the generational approach provides.

Later chapters will look at what 6G might be and then ask whether a new generation is needed or whether intra-generational updates would be sufficient. Many, over the years, and especially with 5G, have argued that it might be the last of the generations.

1.4 Why we will get 6G regardless

While it might not be sensible to use the generational approach for 6G, as later chapters will discuss, it will happen anyway. That "super tanker" has left the port, is steaming to its destination, and very little can stop it. Governments and bodies like the EU have already funded 6G research centres, standardisation work is starting, manufacturers are providing their strong backing, and politicians are already discussing the importance of national leadership. It would be extraordinarily difficult to stop all of this in its tracks, especially as no one entity holds the levers that could do it. The closest would be 3GPP who could decline to develop the standards, but this would not be in their self-interest.

There is a self-interested drive to a new generation that is very difficult to influence, and it may that it takes a few failed generations before stakeholders voluntarily decide to do something different. The eco-system is complex with much informal influence as follows:

- At the heart of the system is **3GPP**, the standards body that coordinates global cellular standards. Within 3GPP the standard gets written and 3GPP publishes it. Timetables for intra-generational and new generations of standards are defined by 3GPP working groups. Without 3GPP, or some equivalent body, there would be no cellular standards.
- All the outputs and key decisions of 3GPP are made by its members – 3GPP has a small secretariat but no employed experts. Any entity can

become a member by simply paying the membership fees. But to have influence, members need to attend many meetings, spread around the world, and produce input papers for those meetings. The papers either need to be persuasive, or the members need to persuade other members to agree and accept them such that their ideas become part of the standards. As a result, being influential in 3GPP is very expensive, requiring many experts, much travel and much work to author papers and manage the politics. The result of this is that only very well-resourced members have substantial influence – for the manufacturers costs are relatively manageable, but for most other players they are material.

- Those members with the greatest voice in 3GPP are the **global manufacturers** – Ericsson, Nokia, Huawei, Samsung and similar. Large Mobile Network Operators (MNOs) can also have a voice although most choose not to be active players since they have other routes of influence that are much less expensive and likely more effective. Hence, broadly, global manufacturers drive standards.
- The manufacturers pursue new generations because they hope that this will result in increased sales as **MNOs** buy equipment and users buy handsets. However, they will only sell equipment if the MNOs choose to buy it. Hence, the MNOs can influence manufacturers by setting out what equipment they would buy and what they would not. But their voice is relatively weak since there are hundreds of MNOs, and less than ten global manufacturers, and MNOs tend not to speak with one voice. For example, the interests of Chinese MNOs can be quite different from those in Europe. Also, MNOs tend to have very few experts, having relinquished nearly all research and development activities long ago.
- Manufacturers try to increase pressure on MNOs to buy their equipment by soliciting support from **politicians** and **academics**. As was clear with 5G, politicians are keen to be associated with new technology and their desire to see strong national deployment makes it harder for the MNOs to decline. Academics provide credibility and publicity to justify the new generation and may also deliver useful research ideas. The funding that the manufacturers and governments

provide is extremely welcome to academics who are often short of resources.
- The cellular eco-system is huge and there are many other interested parties. For example, TowerCos tend to hope that new generations will result in more masts and more deployments on existing masts, benefiting their business. Consultants tend to gain whenever there is change as brought about by new generations.

Broadly, then, new generations are very important, perhaps even existential, for equipment manufacturers. Manufacturers dominate 3GPP in order to drive the standards forwards and, in some cases, can strong-arm MNOs into a position where they feel they must buy equipment, if nothing else to appease local politicians and just in case having the latest generation turns out to be important to consumers.

Critically, there is no counterweight. Nobody in the eco-system has any real interest in preventing a new generation, other than the MNOs if they believe it may be bad business for them. A few independent commentators, such as myself (eg with "The 5G Myth") might speak out, but with a voice many orders of magnitude weaker than that of the supporters. Hence the view expressed above that only through one or more failed generations can the super-tanker of cellular generations be halted as manufacturers realise it does not lead to more sales and politicians and others no longer wish to be associated with what is now seen as failure rather than progress. 5G is now being recognised as a disappointment, as will be discussed in the next chapter, but this is not sufficient to prevent 6G happening.

While we cannot change the fact that we will get a new 6th generation of cellular, we can affect what it sees as the needs of consumers and what it delivers. That is what this book is about.

2 Lessons from 5G

2.1 *A generation like no other*

Generations before 5G aimed to provide better connectivity, fixing the problems occurring with existing solutions. 5G aimed to change the world. Hyperbole around what 5G would achieve abounded. For example, Steve Mollenkopf, CEO of Qualcomm, said[3] "5G will have an impact similar to the introduction of electricity or the car, affecting entire economies and benefiting entire societies". Similarly, O2 Chief Operating Officer Derek McManus said[4] "5G is the most powerful opportunity to strengthen the economy, enrich lives and outperform the global economy. 5G will have a bigger impact than any other technology introduced since electricity." Some said that 5G would be a generation like no other. That has turned out to be true but not in the manner predicted. The expectation was that 5G would lead to vast numbers of connected devices, to new "metaverse-like" ways of communicating, to autonomous cars and robotic surgeons; in short to a science-fiction world.

Instead, 5G has led to increasingly cash-strapped MNOs and a sceptical population. The metaverse, autonomous cars and robotic surgery remain far away.

Articles abound about the failure of 5G. For example, ET Telecoms wrote[5] about how 5G had disappointed pretty much everyone, while in 2023 the Washington Post headlined[6] with "5G was an overhyped technology bust. Let's learn our lesson." Mostly these articles revolve around the ridiculousness of the applications anticipated to emerge, although few seemed to find them ridiculous when 5G hype was in the ascendent.

---

[3] https://www.here.com/learn/blog/5g-infrastructure
[4] https://www.telecoms.com/wireless-networking/we-re-not-convinced-by-the-convergence-hype-o2-ceo
[5] https://telecom.economictimes.indiatimes.com/news/how-5g-disappointed-pretty-much-everybody/98321122
[6] https://www.washingtonpost.com/technology/2023/06/13/5g-didnt-matter/

Perhaps most telling was a paper from one of the poster children of 5G, SK Telecom, who had pioneered 5G in the tech-hungry South Korea. It is worth quoting their paper[7] at length.

> The 5G Vision Recommendation, published in September 2015 by ITU-R, an international standardization organization under the United Nations, triggered the commercialization of 5G services in 2019.
>
> Although various goals have been achieved, we need to prepare for 6G services by checking tasks that have not yet been achieved at this point. A variety of visionary services were expected, but there was a lack of killer service. When 5G was being prepared, services such as autonomous driving, urban air mobility (UAM), XR, hologram, and digital twin were expected. So far, there have been some cases that have not actually led to service activation compared to expectations.
>
> A more objective view regarding the future prospects of 5G as well as the readiness of its surrounding environment could have been helpful in managing the level of expectations for 5G. 3D video, UHD streaming, AR/VR, autonomous driving, and remote surgery are representative examples of 5G use cases that were stated in the 5G Vision Recommendation but have not yet become mainstream. Most of them are the result of a combination of factors such as device form factor constraints, immaturity of device and service technology, low or absent market demand, and policy/regulation issues, rather than a single factor of the lack of 5G performance.

It does not take much reading between the lines to get the message, politically stated, that there was no real demand for many of the services on which 5G was predicated.

It is quite early to declare the failure of 5G. We are now around halfway through the typically decade-long cycle between its initial introduction around 2019 and the anticipated introduction of 6G around 2030. Things may improve

---

[7] https://newsroom-prd-data.s3.ap-northeast-2.amazonaws.com/wp-content/uploads/2023/11/SKT6G-White-PaperEng_v1.0_clean_20231129.pdf

in the second half, and some pin their hopes on the intra-generational upgrades, especially SA, that will bring features needed for some of the more extreme applications. But we need to make decisions on 6G now, based on the evidence of 5G to date.

This chapter looks at what happened with 5G and asks what lessons can be learnt for 6G.

2.2 *Solution in search of a problem*

As set out in the first chapter, a new generation should address a clear problem or business need. That was the case with 2G (delivering capacity and security), 3G (delivering fast data) and 4G (fixing the technical issues with 3G). But there was no clear problem for 5G. Problems need to be general, rather than specific services, since services come and go. Ideally problems should be of the form "more capacity" or "more coverage".

I explored this is some detail in "The End of Telecoms History" where I showed that 4G delivered all that we needed. There is no benefit in faster speeds than 4G delivers and no clear benefits in lower latency. There was a need for more capacity, but this could have been met by more spectrum for 4G. This remains the case – those with a good 4G connection will see no advantage in moving across to 5G and may see disadvantage in higher handset battery drain.

In the absence of clear needs for 5G the industry decided to make some up.

This is not completely outrageous, many successful products and services have been introduced in the belief that users will want them once they experience them, even if they see no current need. This is often known as "build it and they will come" after the film Field of Dreams. But it clearly requires highly skilled judgement based on a deep understanding of an underlying need and all the various factors associated with meeting it, including technology and economics. Steve Jobs was seen as a master of "building it and they will come". Had the 5G community solicited the help of proven visionaries and successful entrepreneurs then making up needs for 5G might not have been as egregious as it turned out to be.

It is not entirely clear where the 5G use cases came from, but it seems likely that it was a mix of academics and manufacturers. Both are terrible places to look for successful entrepreneurship. Neither have strong links into user communities - being at least once removed from customers. Both have strong incentives to come up with the most far-sighted applications that drive the greatest need for research and equipment deployment. Neither have any "skin in the game" should the applications not emerge. Instead, their incentive was to invent applications that justified research and standards directions that they had already set upon.

In fact, many of the successful new ideas for applications come from completely outside the cellular community, from companies developing games, apps and similar. Such companies see little need to engage with 3GPP and indeed are often ideologically opposed to the world of formal standards, preferring and open-source approach.

I commented on the ridiculousness of the proposed 5G use cases in 2016 in "The 5G Myth". They have all proven as ridiculous as I predicted. In 2024, in "The End of Telecoms History" I looked again at the applications to understand whether any were now closer to emerging and concluded not.

Unfortunately, many in the mobile community have decided that the reason the applications have not appeared is because 5G was insufficiently fast, and that "doubling down" on delivering even better performance in 6G will result in these applications finally emerging.

Without the hoped for applications it was a case of "built it and they won't come" – expensive roll-out with no revenue to compensate.

The first lesson is that a new generation should not be a solution in search of a problem. The horse should always be before the cart.

2.3 Poor choice of frequency bands

It is very well known in the radio industry that higher frequencies do not travel as far. As a rule of thumb, doubling the frequency (eg 900MHz to 1800MHz) halves the range. Since the area of a cell is the range squared, then this results in a need for four times as many cells. Cells are very expensive, and MNOs try hard to avoid adding more to their networks.

One of the reasons why 3G disappointed was its use of 2.1GHz compared to the 900MHz that formed the basis of most 2G networks. While MNOs did build some more cells, it was not enough to offset the reduced propagation, and the resulting poor coverage meant 3G under-performed. 4G was introduced at 800MHz, a frequency with slightly better propagation than the 900MHz used in 2G network, meaning no new cells were needed and coverage was good.

The main frequency band for 5G jumped up to 3.5GHz – nearly twice that of 3G and over four times that of 4G (implying 16x more cells). Why did this not raise alarm bells?

The logic for using this band was that to meet the targets for download speeds large bandwidths were necessary. Large bandwidths are only available in higher frequency bands. Hence 3.5GHz. This was, of course, deeply flawed since (1) users had no need or gained no benefit from speeds beyond those of 4G and (2) MNOs could not afford to build 16x more cells than their current networks. But 5G proponents brushed this aside, inventing applications that might require higher speeds and then assuming hugely increased MNO revenues due to all the new use cases.

5G proponents did try to extend its range with the use of beam-forming antennas that delivered additional gain and hence increased range. These worked, somewhat, but required lower frequency uplinks, reducing uplink speed and complicating network design.

The result of these poor choices can be seen in the data from companies such as Opensignal that shows the percentage of time 5G is available to users in many countries is around 10%[8]. Even in the "poster-child" 5G deployment of South Korea, availability ranges from 35%-45%, but in, eg Canada it varies from 8%-10%, in the UK it is 10% across all MNOs, in Germany 9% - 16%

---

[8]   https://www.opensignal.com/2023/06/30/benchmarking-the-global-5g-experience-june-2023

and so on. Saudi Arabia is better at 18%-20% but has long had policies to promote 5G deployment. Opensignal only lists six operators globally that have more than 30% 5G availability.

These numbers do need to be taken with some care. They do not imply that users can only connect to 5G for 10% of the time, but that the network has only chosen to connect them for that percentage of time. In some cases, there will be 5G coverage but the network keeps the user on 4G because there is no need for the 5G capabilities and 4G is more efficient. In the few cases where operators have 5G SA, such as T-Mobile in the US, then numbers are higher as devices connect directly to 5G rather than indirectly through 4G control channels.

What we can read into these numbers is that most of the time users either cannot be connected to 5G or do not need to be connected to 5G.

This approach is likely logical for MNOs who use 5G more to enhance capacity than to deliver new services. Spending more either on coverage or in moving to SA might increase the percentage 5G availability, but the consumer experience would be unchanged and hence revenues would not increase.

Because of these factors, the current typical 5G availability of 10-20%, which has taken five years of 5G deployment to reach, is unlikely to change much. It seems likely that most will never see 5G availability for more than a quarter of the time. Because 4G provides all they need they likely will not care.

But there are some important implications.

- Firstly, the incentive to develop any apps that require 5G performance (eg very high data rates or very low latency) is low. Customers are not going to be happy with an app that works only some of the time. And with no apps that need 5G the incentive to deploy it further is low, feeding a vicious circle.
- Secondly, the availability is heavily influenced by the frequency band. 3.5GHz has materially worse propagation than bands used for prior generations. And yet 6G proponents are looking at even higher bands

in the region 7-14GHz. This could result in 6G availability remaining in single figure percentages for its entire lifetime.

The second lesson is to stick to frequency bands that work with the current cellular grid structure – which ideally means below 2GHz.

2.4 Politics and the race to 5G

The involvement of politicians in 5G did not help – or at least it helped the manufacturers but not others. Noting that 4G had delivered tangible benefits to consumers, politicians decided that being associated with 5G would deliver personal benefit. It showed them as being forward thinkers, associated with leading edge technology. Perhaps some genuinely thought that 5G would be important to their country and that without intervention it would not materialise, but if so, they had not given the matter much thought. If 5G was such a good thing then there would be no need for intervention since MNOs would be keen to deploy as fast as possible to realise the revenue that would result.

Many politicians saw 5G as a race which their country had to win otherwise dire consequences would occur. For example, Brookings wrote[9] "The United States and China are in a race to deploy fifth-generation, or 5G, wireless networks, and the country that dominates will lead in standard-setting, patents, and the global supply chain". Clearly the concept of a race was false – those countries "behind" in 5G deployments are not suffering any adverse consequences and tend to have lower costs for mobile service.

Politicians have a powerful voice, and their echoing of the hype from the manufacturers as to how wonderful 5G was going to be meant much greater exposure of messaging about 5G to the wider population who were told that 5G would transform their lives.

This made it very hard for MNOs not to put a positive spin on 5G and not to deploy it. To be naysayers might incur political ill-will with potentially poor outcomes in matters such as regulatory decisions and grants, and with the

---

[9] https://www.brookings.edu/articles/navigating-the-us-china-5g-competition/

population being "educated" on the importance of 5G it might result in subscribers moving to those networks that had the fastest or most extensive 5G deployment. Eventually, of course, subscribers would realise that there were no benefits from being on the "best" 5G network, but this could take years, and much damage could be done in the meantime.

There are areas where politicians should get involved in cellular, and these are where MNO economics do not lead to outcomes that are optimal at a country-level. Issues like rural coverage and network reliability fall into these areas. But next generation mobile deployment should not.

It is hard to derail a politician, but had MNOs briefed them early in the development process on the challenges with 5G and had MNOs worked on a more likely set of use cases, or shown how those put forwards were ill-advised, then perhaps some of the political intervention might have been prevented.

But broadly politicians like to be associated with success. Since 4G was better than 3G, then an association with 5G was bound to be seen as advantageous. Now that 5G is seen by many consumers as a disappointment, a political association with 6G may not be as appealing.

The third lesson is to engage politicians early to try to ensure that any intervention is helpful.

2.5 Expenses imposed on MNOs

5G was much more expensive, at least on an equivalent coverage basis, than 4G. With 4G MNOs could broadly just upgrade existing base stations, add new, but similar size antennas, and leave much else unchanged. 4G was quickly deployed across entire networks. A 5G site upgrade requires much larger and heavier multi-input multiple-output (MIMO) antennas which are often difficult to accommodate at existing sites. Because the range of 5G is lower it ideally requires a lot more base stations. Should the envisaged applications have emerged these would typically have required further investment in delivering complex solutions and perhaps undertaking tasks such as network slicing.

In most countries, as 5G became available, MNOs decided to maintain existing levels of capital expenditure (capex). Instead of spending on 4G coverage expansion and maintenance they spent on 5G deployment. Because of the higher costs of 5G this has meant a much slower deployment of 5G to much less of the population than for 4G. The result was discussed above: a typical 50% 5G population coverage across many countries (compared to a 99% 4G availability) and a much lower real-world availability. Many MNOs are now saying that they cannot afford a similar experience for 6G and that any 6G upgrade must be software only (and hence avoiding the expense of base station upgrade).

The lesson here is that any costs imposed on MNOs must be affordable and have a positive business case.

2.6 A hype machine in overdrive

The hype surrounding 5G drowned out any dissention and indeed made dissenters look like naysayers. It prevented any sensible discussion.

Of course, there is a need for enthusiasm over new ideas and the Gartner hype curve charts how many new concepts are over-hyped (resulting in disillusion). To say that new generations should avoid all hype would be unrealistic and some previous generations such as 3G have had plenty of hype. But balanced discussion should be possible, and journalists and academics should look critically for both sides of an argument.

The lesson is that critical voices should be encouraged and balanced discussions held in the interest of countering inevitable hype from those with most to gain from the next generation.

2.7 Is there any hope for 5G?

There are still new features to be deployed in 5G, centred around SA and including lower latency and network slicing. These might enable new applications. Notably, 3G did not see its vision of mobile data succeed until 2007 – three quarters of the way through its decade – so it may be too soon to judge 5G.

But the signs are not good. If most users only access 5G for 10% of the time, then there is little reason to offer them applications dependent on 5G. And without new applications this percentage will not grow to the 90%+ level needed to make most new services attractive.

Private networks do not face these issues since coverage can be tailored to the campus or factory, but such networks will be relatively small in number – certainly insufficient to justify all the investment in 5G. This is because most needs are met by wired machines or by Wi-Fi, and because in most developed countries there are relatively few factories, such that the value of the market for private cellular networks is a tiny fraction of that for public networks.

2.8 Implications for 6G

The over-riding lesson for 6G is to find solutions to problems, not the other way round. And there are plenty of problems, as will be discussed in the next chapter.

Other lessons are:

- Stick to frequency bands below 2GHz. If this is not possible look closely at the economics.
- Have a strong MNO voice setting out their perception of need and educating politicians in a balanced way, which may require MNOs to recruit skilled individuals who are able to do this.
- Understand the economics of 6G as much as the technologies.
- Encourage a balanced debate with independent voices, especially those with proven track record.

The next chapter looks at the problems that are clearly visible with current mobile networks and that 6G should focus on tackling.

3 The problems that are clearly visible

3.1 Ask and you will be told

To find out what consumers want from mobile communications all you need to do is ask them.

Published surveys of consumer preferences are surprisingly hard to find likely because most remain private. Virgin Media recently discussed one[10] it had commissioned, saying:

> A recent report from mobile network benchmarking specialist GWS revealed that the UK's major mobile operators are hitting the 'sweet spot', with consumers satisfied with their everyday mobile network speeds ranging between 1-5Mbps. Even more importantly, the report found that speed is not a key driver for people to move networks, with poor signal and blackspots ranking much higher than speed as a reason to change.

Opensignal published[11] the results of a survey of US Consumers, as shown in Figure 3-1.

As the Opensignal survey shows, cost is the single biggest factor (21%). But the next three largest – reliability, network quality and coverage - are all part of the same need, to always have connectivity sufficient to undertake whatever task is required. As the GWS survey reports, this is around 1-5Mbits/s of consistently available connectivity. Collectively, reliability, quality and coverage add to 54% of the consumer responses. Everything else pales in comparison.

---

[10] https://news.virginmediao2.co.uk/leaving-the-vanity-metrics-behind-and-focusing-on-what-matters-customer-experience/
[11] https://www.opensignal.com/2024/02/08/the-opensignal-global-reliability-experience-report

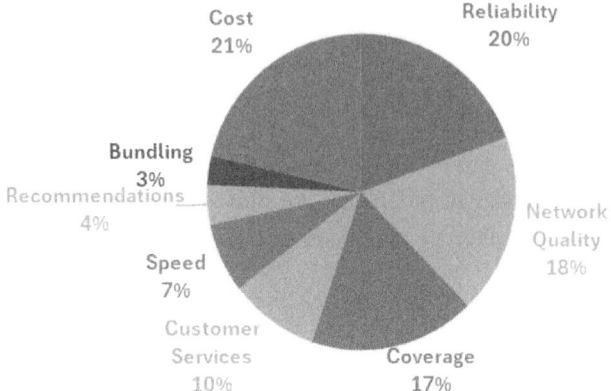

Figure 3-1 : Opensignal user survey: Source Opensignal

This should not be surprising. We are all highly reliant on our phones for so much in our lives, not just communication but navigation, payment, entertainment and more. When phones do not work, even momentarily, it is a problem, especially if unexpected (we tend to get used to not-spots on our daily travel routines and work around them). Above all, we want a phone that always works. The fact that being always connected features so highly tells us that it is rarely delivered.

Hence, ensuring subscribers are always connected with sufficient quality to enable all typical applications at a low cost is a clear problem to resolve. Part of this is about having sufficient capacity and 5G has enabled networks to grow their capacity in line with demand growth. But another part is coverage and the 5G standard has inferior coverage to 4G at a greater cost.

Delivering faster networks with more capabilities is not important. The UK Communications and Market Authority (CMA), in considering the proposed Vodafone-Three merger said[12]:

---

[12] https://assets.publishing.service.gov.uk/media/66e3ca8a0d913026165c3df4/Summary_of_provisional_findings.pdf

For example, most consumers told us that they would not be willing to pay more for better quality (with 76% unwilling to pay for faster speed, and 59% unwilling to pay more for more reliability). Ofcom also noted the current limited evidence of customer willingness to pay a premium for services that rely on 5G SA capabilities.

Consumers want connectivity everywhere they go at current service levels, not enhanced service levels.

To be clear, this does not mean ubiquity. There is no point having coverage in places where nobody goes. There will likely be a lower threshold of "footfall" below which connectivity makes little sense. We should ensure that *almost* everyone is connected *almost* all of the time. Targets need to be pragmatic as much as they need to be challenging.

3.2 Ensuring users are always connected

3.2.1 Introduction

There are many situations in which connectivity may be insufficient:

- The most obvious are large areas without coverage, typically in rural parts of the country. While large geographically, these tend to affect relatively few people because of the low population density in the region.
- There are not-spots in urban areas, typically smaller areas where there is building shadowing, or no cells close by. These can tend to affect large numbers of people but generally for short periods of time (unless they become stationary in the not-spot).
- In-building coverage can be challenging, especially where coverage is provided from outdoors and must get through the building fabric.
- There are connectivity challenges in trains and underground metros.
- There are areas with coverage but insufficient capacity such that the quality of connection suffers.

Coverage will differ between MNOs, such that in some cases not all subscribers will be affected.

Solutions exist for all these challenges, the reason that they have not been used is generally economic – the MNOs cannot see a viable business case in making the investment. This is because they do not expect to see increased revenue as a result. This may seem strange - if users value being always connected so highly why will they not pay more for it or move to the MNO that offers the best connectivity, creating competition between MNOs on the basis of quality of coverage? Broadly this is because consumers find it very difficult to compare the coverage of different mobile operators. A consumer would like to know which operator would give them the best coverage across the areas where they live, travel and work but the only way they can reliably compare operators at present is to have multiple phones and subscriptions and constantly compare them.

The solution is relatively straightforward. Consumers could download an app that (with their permission) tracks where they go and when they use their phone. Aggregated over a week or two this would provide a sound basis for the consumer's need. Their usage can then be correlated against crowd-sourced coverage maps from each operator giving the coverage and speed of connection across the locations they visit. From this the optimal operator can be determined.

This is beneficial for the consumer who can select the best operator for them, for the country because consumers are better connected, and for coverage enhancement as operators are now incentivised to improve their coverage to avoid loss of customers.

Delivering such an app is relatively straightforward once a crowd-sourced solution is in place that can provide the underlying coverage data needed. This raises the subsequent question as to why such an app is not available, and the answer again is economics. The app would have to be provided by an independent entity (since nobody would trust one operator to produce an app that might tell their customers to go to a different operator) but there is no obvious way for the app provider to monetise their app – users do not expect to pay for such comparison advice. It would make sense for governments to fund it since the improved coverage that results would deliver citizen benefits for the country, but few governments are sufficiently enlightened to do this.

The table below shows the potential solutions to each of the coverage challenges listed above. These are then discussed in more detail in the remainder of this section.

Challenge	Solution(s)
Urban not-spots	Small cells providing in-fill, network roaming to other networks with coverage
Rural gaps	More cells, high altitude platforms such as aerostats and drones, satellite direct-to-device coverage
In-building	Use of in-building Wi-Fi, dedicated in building cellular networks where demand merits the cost
Travel	Tailored solutions for trains based on microwave links to carriage roofs and in-train Wi-Fi, leaky feeder in tunnels
Capacity issues	Capacity expansion with 5G, additional spectrum or cell splitting, possible demand throttling

Table 3-1 : Solutions to coverage challenges

The aim of the following subsections is not to provide great levels of detail on the possible solutions – there are other resources that already do this – but to bring out the most likely solutions and the implications for any 6G standard.

3.2.2 Urban connectivity

There are still many not-spots in urban areas, typically caused by shadowing from buildings and the challenges of finding optimal cell sites. Often these not-spots are MNO specific – other MNOs with different cell sites and different frequency bands will have not-spots in different locations.

Unlike rural areas, coverage from the skies is unlikely to be a viable solution. Satellite reception in urban areas is harder because of the reduced visibility of the sky and satellites will likely not be able to provide the capacity needed. The rapid handovers to and from the terrestrial network will make it hard to use satellite solutions. HAPs is more workable but the economics of using a HAPs platform to cover urban not-spots is more challenging as is the ability to dedicate spectrum to HAPs when all the spectrum is likely already being used for terrestrial transmission.

This leaves two avenues:

- Infill using small cells.
- Roaming across MNOs.

Infill is the "classic" solution. Operators are slowly adding more small cells in cities, albeit reluctantly as the economics of small cells are poor. Small cells are easier to site than macrocells, and often can be located in places that in-fill not-spots. The only issue is the cost. Filling hundreds of small not-spots in a city might require hundreds of small cells, and there is little upside to the MNO (because consumers rarely switch because of coverage, as discussed earlier).

"National roaming" – roaming to a different MNO when in the home country – is much less expensive. Collectively the MNOs may have very few urban not-spots and can address these jointly. The roaming itself is almost cost-free - it is already a feature of mobile networks. However, it raises many policy issues and tends to be strongly resisted by MNOs. Broadly, the MNO with the best coverage has the least to gain and the one with the worst coverage the most. MNOs also suggest it disincentivises further investment in coverage since one MNO's investment immediately becomes available to all other MNOs. However, as individual coverage investment (as opposed to shared investment) is now rare, this has become less of an issue.

As with all policy and commercial issues there are solutions to national roaming concerns. For example, a government or regulatory obligation to fix not-spots (eg in return for licence renewal) could force MNOs to consider the lowest-cost ways to achieve this. MNOs might collectively be able to agree compensatory payments from the MNOs with the worst coverage to those with the best, or similar. Roaming could be restricted to urban not-spots, such that overall coverage differentials across the country were maintained.

There are also technical issues with roaming that would need to be resolved in a 6G standard, for this approach to work well:

- The process of moving from one network to another is slow, often taking minutes rather than seconds. Activities like authentication can take a long time as can the process of detaching from one network and attaching to another.
- Any active IP sessions will be torn down in the process, dropping all calls, video streams, etc.
- Handsets on roamed networks are set to frequently search for the home network, consuming battery power in the process.

All of these could be solved with an over-arching coordinating function sitting above the multiple cellular networks, handling aspects such as authentication and maintaining IP sessions. It could also provide geofencing data to handsets when roaming, enabling them to only search for their home network when it has become available.

Implications for 6G

The implications for 6G are:

- Enable small cells to be implemented at the lowest possible cost.
- Resolve the current issues with national roaming that prevent near-seamless service and drain battery power.
- Facilitate complex and flexible national roaming approaches, for example that can be tailored to each specific not-spot.

3.2.3 In-building connectivity

In-building connectivity is likely to be the largest source of irritation for consumers. We spend much of our time in buildings and tend to use devices more when stationary indoors than when outdoors. But building facades cause a "penetration loss" for radio signals entering the building from outside. This can be considerable, resulting in poor quality or lack of connectivity. Many of us are used to moving closer to the window to make mobile devices work.

Even where devices work well, they can impose substantial load on the network since the reduced signal level causes devices to utilise much more of the network resource[13].

The solution is completely obvious – coverage needs to be delivered from transmitters within the building. This avoids the penetration loss and indeed, the building facade now shields users from interference and reduces interference into the external network. However, putting a cellular transmitter in every building is impractical. Various attempts to do this over the years, including femtocells and similar, have all failed. There are some buildings with cellular coverage, such as airports, stadiums and shopping malls, but these are restricted to high footfall public premises. It seems very unlikely that the number of buildings with dedicated in-building cellular coverage will materially change.

But almost every building already has radio coverage – from Wi-Fi. It is vastly cheaper, much more practical, and much quicker to use this connectivity than seek improved cellular solutions. Indeed, we all tend to use Wi-Fi when at home or in the office, or other much-visited locations where signing onto the Wi-Fi is considered worthwhile. Cellular systems implement voice over Wi-Fi which allows voice calls to be made and received[14] (although increasingly we make use of applications such as WhatsApp for voice calls rather than rely on cellular). The only issue is extending this "sign on" to all buildings, or at least all commercial buildings since residential dwellings tend to be less of an issue with lower penetration loss and a simple way to ask the host for Wi-Fi access if needed.

The way to resolve this is a single sign-on for all commercial Wi-Fi networks that avoids the need to proactively select the network and enter the password. There are effective technical solutions to this and there have been good examples

---

[13] This is because at a lower signal level the bits transmitted per unit of radio spectrum reduces, causing transmissions to take longer. A mobile with a poor signal could require ten times the network resource of one close to the base station to undertake the same task.

[14] In practice, cellular voice over Wi-Fi is not a good way ahead since the cellular network effectively tries to control the Wi-Fi network, demanding better quality for voice packets and tunnelling packets into its own network. This is blocked by some Wi-Fi networks and may not work when roaming. True OTT services, such as WhatsApp are better suited to a multi-network world.

of cellular mobile operators creating coverage partnerships with commercial building owners. But not yet at scale, although companies like American Bandwidth[15] are making good progress.

Why would commercial (and perhaps private) Wi-Fi owners agree to this? Commercial Wi-Fi is often provided as a service, to hotel guests, those in coffee shops, visitors to corporate buildings and so on. These entities have an overhead associated with communicating the Wi-Fi ID and password as well as meeting various legal requirements such as preventing access to certain materials and recording identities of users. Users may prefer to go to venues where they have already logged onto the Wi-Fi rather than having to go through a process of logging into a new Wi-Fi network. An automated process, ideally where legal requirements are managed by the network aggregator, could save cost for venue providers, while making their underlying service (eg a hotel room) more attractive. Already, for example, coffee chains outsource their Wi-Fi to a company such as The Cloud[16] and likely pay them for this service. If the universal sign-in gained momentum then those who were not part of it might experience customer frustration, never a good thing.

This is not a new idea, and there have been many attempts in the past, including the still-existing BT OpenZone arrangement and university EduRoam scheme. They have not gained traction partly because the technology was not there and partly because MNOs had other priorities. Technology has improved, especially with the recent OpenRoaming initiative. This may be an idea whose time has finally come.

Achieving single sign-on for all commercial Wi-Fi networks might require government intervention to establish the basic "club" with enough scale and momentum to encourage others to join. For example, European governments could open their EduRoam shared Wi-Fi to all users and allow other entities to add their Wi-Fi resources into the scheme. More complex mechanisms that allow those volunteering their Wi-Fi to get some recompense could be imagined.

---

[15] https://ameriband.com/
[16] https://www.sky.com/wifi

Such a solution might be thought to require integration between cellular and Wi-Fi. As will be discussed in subsequent chapters, whether this is the case is unclear. Currently there is no integration, yet handsets manage the process seamlessly, linking to both Wi-Fi and cellular networks and preferring one over the other according to internal rules. Better integration might help with authentication and routing of incoming calls and data and could potentially deliver better handover decisions between networks.

Implications for 6G

The implications for 6G are:

- 6G should not aim for another solution to indoor coverage but instead assume that it can be delivered via Wi-Fi (except in a few special cases where private cellular networks have benefit).
- 6G should embrace Wi-Fi, with it being very much part of the overall solution. There should be integration between 6G and Wi-Fi, and 6G investment should be minimised where Wi-Fi can be leveraged. Indeed, and this is a point returned to later, the 6G standard might sit above both cellular and Wi-Fi as a unifying approach, rather than applying only to cellular.

3.2.4 Rural connectivity

While many developed countries have reached population coverage levels of 98% or more, it is normal for geographical coverage to be 90% or less. While 100% rural coverage is likely unnecessary and overly expensive, much higher levels of coverage are needed to deliver connectivity where there is significant footfall. Many countries have placed requirements on operators to extend coverage, for example in return for "free" licence renewal, and others have implemented jointly funded "shared rural networks" (SRNs) where government part-funds the deployment of increased coverage.

All these approaches are "classical" – they stay with the standard technique of expanding coverage by building more masts. While this undoubtedly works, it is slow and expensive. Many schemes, such as the UK's SRN, are running behind schedule, primarily because it is difficult to find mast sites in rural areas. This is because:

- Many rural areas have planning restrictions since they are considered areas of outstanding national beauty or similar.
- Many areas are hilly or mountainous, making the mast location critical – needing to be on top of a ridge or hill.
- Getting to the site to prepare the ground, install the mast and then kit the site out can be very difficult and often requires helicopters to fly in equipment and, in some cases, the building of access roads.
- Rural sites rarely have power or communications available, and providing power can be very expensive.

Whether coverage wherever people go can ever be achieved by this approach is unclear. It is certainly a slow and expensive way ahead.

The alternative is "coverage from the skies". This encompasses two types of solution – high altitude platforms (HAPs) and satellite direct-to-device (D2D). Both have the merit of avoiding the need for masts.

There is much scepticism about HAPs, based on decades of failures. In recent years, for example, the much-discussed Google Loon project of free-flying balloons was withdrawn due to impracticality. Many current projects are behind schedule. However, there are some new solutions, learning lessons from previous failures, that appear sensible and practical. For example, tethered aerostats stay in the same location, can withstand severe weather (or be winched down in extreme cases), can send power and communications via the tether and hence deliver huge capacity. It is worth reassessing some of the latest HAPs concepts, albeit with a healthy dose of scepticism based on experience.

There are many variants of coverage from the skies that provide a range of possible coverage levels as summarised below. *Note: this is an area where there is limited data available, where innovation is leading to improvements and where costs are rarely available, so these are my "back of the envelope" estimates.*

	Range per device	QoS and capabilities	Cost for UK rural coverage
Tethered aerostats	~40km	Full 5G capabilities and fixed wireless access (FWA)	£0.5-£2bn
Drones	~100km	Limited to 2G services	<£1bn
Aircraft	~100km	Full 5G capabilities, some FWA	Unclear, likely similar to aerostats
Basic low earth orbiting (LEO) constellation	~1000km	Emergency messaging only	Around £100m if global coverage provided
Enhanced LEO	~1000km	2G level services	Unclear, perhaps double basic LEO

Table 3-2 : Coverage from the skies options

I have tried to provide cost estimates for the UK as an example to allow comparison between the options. The costs will differ for other countries, for example satellite costs are independent of the area to be covered (since satellites provide near-global coverage anyway) whereas HAPs costs tend to scale with the area to be covered.

Of course, each of these has its challenges. Tethered aerostats still need "sites" although the location of these is far less critical than for masts – they can be, for example, outside national parks and located in areas away from population. With the much greater coverage radius only around 1%-2% of the number of terrestrial sites are needed. Satellite networks need to be launched (as some have been) and there are many issues associated with spectrum and similar discussed below. Aircraft need to be flown and maintained. However, the advantages appear to be overwhelming. Coverage from the skies is much less expensive than terrestrial, much faster to deliver, and some solutions can provide excellent service levels.

There is no need to choose between HAPs and satellite – both can be used. HAPs can provide coverage to areas where there is significant population density or footfall, and satellite to areas where footfall is occasional. Users would transfer from terrestrial networks to HAPs and then to satellite as they move to increasingly remote areas.

This approach requires HAPs and satellite solutions to be integrated in some manner with the cellular network. This integration is already somewhat underway with the development of the non-terrestrial network (NTN) standard within 5G. This essentially adapts 5G to work over links with greater delays (primarily for satellite). HAPs solutions might be able to work with standard 5G, especially the lower altitude solutions where the distance to the "base station" is well within 5G terrestrial capabilities. Not all satellite solutions embrace NTN. The Apple iPhone solution is proprietary, and the "integration" happens in the phone in a very simple manner – if there is no cellular connection then the handset makes a satellite connection available for certain emergency-type services. The Starlink solution appears to use standard 4G rather than 5G NTN and it is unclear how this is integrated into a network, although since Starlink partners with MNOs, they can work this out between them.

D2D challenges

There are other substantial questions surrounding D2D. The reason that we only have very limited D2D services today is the difficulty of receiving a signal transmitted from a satellite on a conventional handset. The problem is that any signal will be very weak. Satellites only have a limited amount of power to use for transmitting, as do handsets, and the distance to a satellite is huge compared to a normal cell tower. Conventional satellite communication systems overcome this challenge with high-gain antennas at the receiver such as the small dishes of VSAT receivers or the bulky antenna of the Iridium handset.

Performance can be improved through using more bandwidth. The fundamentals of physics tell us that bandwidth can be traded for power. Using a wider bandwidth channel allows lower power levels to work. Keeping the data rate to a low level which would only need a tiny bandwidth with a terrestrial service and then using a much wider bandwidth to the satellite, can allow the "link budget" to work. But the downside is a very limited service.

There is little spectrum available for satellite communications, in turn limiting what can be achieved with greater bandwidth. To date, at a global level, spectrum has been divided into various uses, one of which is satellite. Because satellite signals are so weak, no other use can be tolerated in most satellite bands. Hence, the ITU designates some spectrum as being for satellites. Within this designation, there is a sub-designation for the "mobile satellite service" (MSS) used by companies like Iridium and Globalstar, the satellite operator that Apple has partnered with. MSS bands have the benefits of global, interference-free operation but there are few MSS bands, and they do not have much bandwidth available. Using MSS spectrum also requires handsets to be modified since it is a different set of frequencies than standard mobile use. This was viable for Apple since they built the handset and designed the service, but harder for other satellite operators that do not build cellular handsets - as Iridium found.

A different option, as proposed by SpaceX, is to use spectrum dedicated to mobile use – the mobile service (MS) spectrum. This has two advantages – massively more bandwidth than MSS spectrum and handsets can already tune to these bands. But it comes with big challenges – the spectrum is already used by mobile operators who will not want interference from satellite transmissions, and the international regulatory environment does not currently allow for satellite transmission in these bands.

Hence, there are currently two camps for those planning to deliver D2D services. In the MSS camp is Apple, Iridium and other established operators such as Viasat. In the MS camp are the newer entrants such as SpaceX and AST.

It looks very hard to deliver both ubiquity and full service. Ubiquity tends to require MSS spectrum which does not have the bandwidth for a full service. A full service requires the bandwidth of MS spectrum which has challenges in being ubiquitous.

The main problem with MS spectrum is that it is typically intensively used for mobile communications. The figures below show the challenge. Terrestrial cells are around 3-5km in diameter in the rural areas where coverage may be sporadic. Satellite cells are much larger, often around 50km across, depending on the size of the array on the satellite.

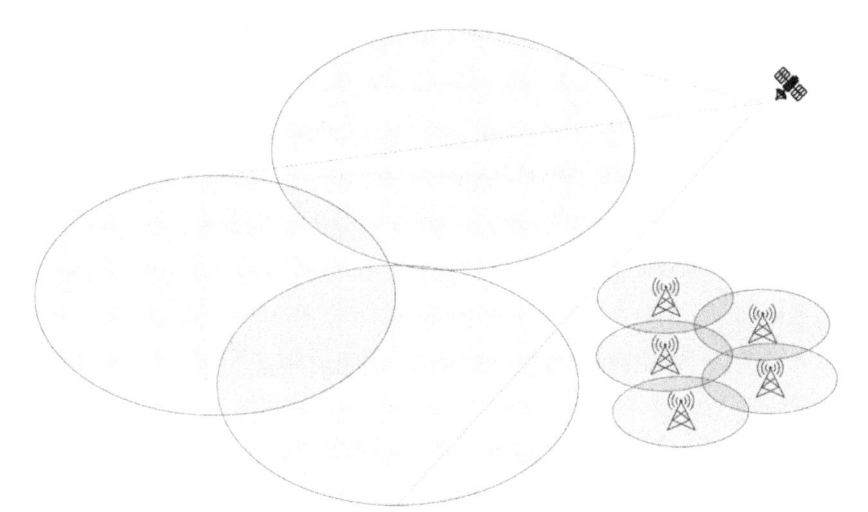

Figure 3-2 : D2D, simple situation

There needs to be a gap between terrestrial usage and satellite use to avoid interference so as the figure above shows there will be areas of no coverage, and these may be quite large where it is difficult to fit the satellite beam pattern around terrestrial coverage. Because the satellites in LEO orbit move quickly across the sky these gaps will change location and size every few minutes making coverage hard to predict.

In a situation such as the one below it could be nearly impossible to deliver coverage since the gaps between the terrestrial coverage are smaller than the satellite beams. This could be the situation in many countries, such as in Europe, where there are not large contiguous deeply rural areas.

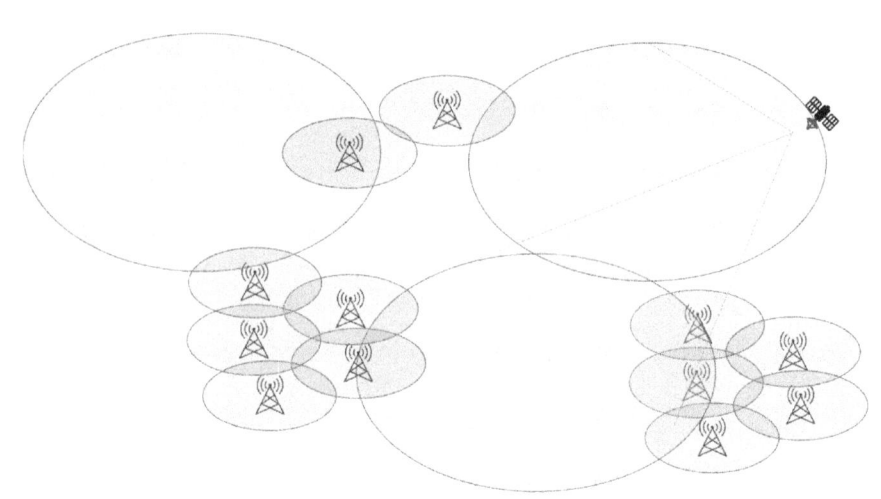

Figure 3-3 : D2D, more complex situation

If the mobile operators are prepared to frequency-plan with the satellite operators, then it may be possible to do better. The figure below shows a situation where in the dense areas both low and high frequencies are used with lighter terrestrial cells representing low frequencies (eg 800MHz) and darker representing both low and high frequencies (eg 2.5GHz).

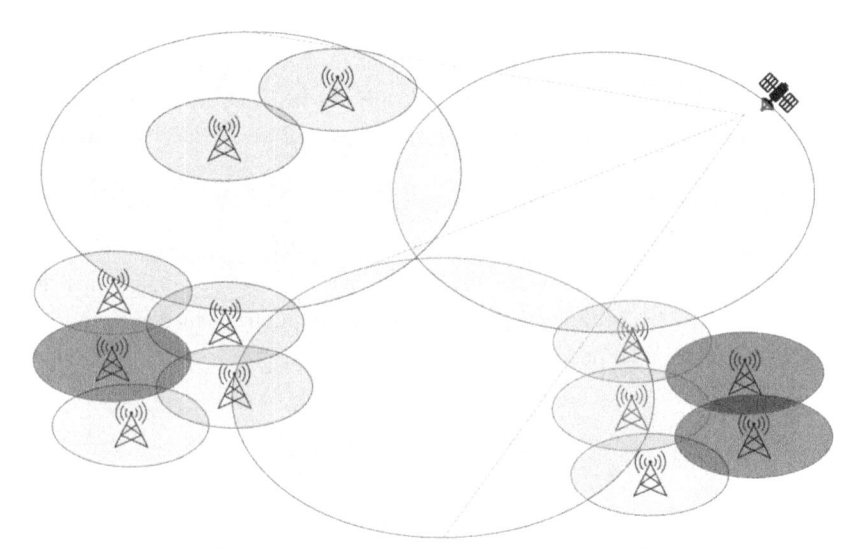

Figure 3-4 : D2D, multi-frequency scenario

The problems are clearly visible

If the high frequencies are used by the satellite, then its transmissions can overlap the low frequency-only areas without causing interference and potentially without leaving gaps. This assumes the beams can be dynamically steered as the satellite rapidly moves overhead.

The situation becomes more challenging when national borders are involved. Satellites will need to leave guard bands near the borders unless they have a similar arrangement with the regulator and the operators on the other side of the border.

This suggests that the service might work best in countries with large rural areas and few borders – the US, Canada, Australia and a few other similar countries. It shows just how difficult ubiquitous coverage will be using MS spectrum.

The ITU sets out what each band of spectrum can be used for. Few of the MS bands have any allocation for satellite usage. Countries can choose to depart from the ITU regulations if their usage does not cause any interference into other countries – this is the position that the US has taken with its proposed "supplementary coverage from space" (SCS) regulation.

However, while regulators in the more advanced countries may be willing to depart from ITU regulations, those in smaller or less developed countries generally are not. They will often not have the staff or expertise to assess whether departing from ITU regulations will cause issues and may prefer to "play it safe". These countries are also likely to be the ones with the weakest coverage and hence where satellite fill-in is most important. This implies that for global coverage using MS spectrum to occur, an ITU designation for satellite use in mobile bands will be required.

An ITU designation can only be made at a World Radio Conference (WRC), the next one is scheduled for 2027. It is far from certain that all the changes needed will be made at WRC-27 and it might take until WRC-31 for all the appropriate mobile bands to have satellite allocations throughout the world.

Considering the argument above a ubiquitous and full service is near-impossible. Two years after Apple launched their service, which is minimal in terms of capability, it was still only available in 16 countries[17]. The SpaceX proposal only has regulatory approval in the US and its service level is unclear.

In summary, D2D's alluring promise of ubiquitous, fully featured global connectivity is not likely to be realised any time soon, if ever, but a reduced service level in a few countries may be sufficient to justify launching suitable satellites.

Implications for 6G

Expanding rural coverage from the skies is the best way to resolve rural coverage issues for many countries. Most HAPs solutions can integrate easily and directly into mobile networks, but satellite systems might require more integration, especially concerning spectrum. Much of this may well be resolved during the 5G era, but 6G could enable better outcomes. For this 6G would need to:

- Enable network sharing, for example across SRNs which might comprise HAPs, in flexible and seamless ways.
- Allow for satellite D2D use of spectrum to work efficiently with terrestrial use, perhaps implementing dual-connectivity approaches and similar.
- Find ways for satellite to work efficiently with cellular, for example minimising paging and location traffic over satellite while maintaining contractability.
- Ideally put in place a mechanism that allowed satellite spectrum use to coordinate globally with terrestrial use.

3.2.5 Travel connectivity

In many countries connectivity on trains is poor, with not-spots in tunnels and cuttings and networks unable to deliver the capacity needed within a carriage where many people who, often having little else to do, are inclined to heavily use mobile connectivity[18].

---

[17] https://support.apple.com/en-gb/101573
[18] Gone are the days when passengers used to read newspapers or even books!

The ideal rail solution is one that can deliver gigabits of data to trains using dedicated mmWave radio solutions to antennas on the carriage roof which are linked to Wi-Fi and perhaps cellular mini-base stations inside the carriage. This can easily be deployed on mainlines which have overhead power where the wireless solution can use the same gantries and can be deployed on other lines using dedicated small masts for the mmWave radios.

This would be a shared network – it would be cumbersome and expensive to build multiple radio solutions on the same lines. It is generally better implemented by a dedicated entity with links to the railway companies to facilitate access and manage safety concerns. Funding would typically need to be provided by government since the MNOs would likely see little revenue benefit and the railway companies generally will have limited incentives to enhance connectivity.

Implications for 6G

The implications for 6G are minimal other than providing strong support for shared networks.

3.2.6 Capacity issues

The final challenge in ensuring users are always connected at good quality is areas where there is good coverage but insufficient capacity. These are often very high-density areas such as train stations, but can occur almost anywhere, especially when there are unusual events such as festivals or traffic accidents.

This will become less of an issue in the future as existing capacity challenges are met because, as set out in "The End of Telecoms History", data growth is slowing and will soon plateau. Hence, congestion will not increase further in future.

The solutions to congestion are:

- Build more capacity.
- Off-load more traffic to other networks.

All of this is well-trodden ground. More capacity can be delivered through using more spectrum, increased cell density or re-farming to a higher generation (eg 3G to 4G) which has increased spectrum efficiency.

Off-loading is less used, but if the improved access to Wi-Fi, discussed above, were implemented then this would off-load substantial amounts of traffic from the cellular network, and likely resolve most capacity issues.

Also worth a mention is national roaming for efficiency. Instead of roaming only when there is no home network coverage, a user could roam to another network where the signal from its base station was stronger than that from the home network (and vice versa). Because this reduces the times when a handset has poor signal, which, as explained above, result in high network loading, it can dramatically reduce congestion as well as enhancing the user experience.

Implications for 6G

The implications for 6G are minimal since data growth is near its end (which, as we will see, is in stark contrast to one of the objectives of 5G and 6G of enabling dramatic growth in network capacity). Better support for off-loading to Wi-Fi and for national roaming, already both mentioned earlier, would result in a low-cost and rapid reduction in loading and congestion.

3.2.7   A role for neutral hosts?

In some cases, cellular networks are deployed and operated by third parties – entities other than the operators. Examples of this include metro (underground) networks and private networks on campus and in public buildings. At present, for this to work, the operators give the neutral host the rights to use their spectrum (or in some cases the neutral host uses shared spectrum where available) and the neutral host sends traffic to the appropriate core network using an approach known as multi-operator core network (MOCN).

Neutral hosts are appropriate where:

1. Deployment is uneconomic for any one operator but could be economic if all operators share a single network.

2. Deployment is logistically very difficult (eg in underground tunnels) and it is only practical to deploy a single network for all to share.
3. Coverage is required over a private area (eg a factory or port), operators will not provide what the owner needs, and so a self-deployed system is required, but access to the public is seen as beneficial. This is often termed a "private network".

Neutral hosts blend somewhat with shared networks. For example, all the operators could collectively share infrastructure including active equipment and use MOCN in a jointly funded vehicle that is functionally identical to a neutral host but differs in ownership.

Neutral hosts could be a way to deliver improved coverage without the operators having to pay capex and in a cost-effective manner through network sharing. But it is hard to see how this makes a difference in the bigger picture. For example, a neutral host could deploy HAPs coverage to rural areas, enabling all operators to access using MOCN. Equally, the MNOs could do this collectively, in the manner that they are implementing the SRN in the UK. All that the neutral host operator achieves is turning operator capex into opex (in the form of the on-going "rental" fees). But operators ought to be able to change between capex and opex in the financial markets as needed. If the operators cannot readily work together (eg because of personalities) then the neutral host may be a way to deconflict relationships.

If private networks are being deployed, then enabling access to these by the public would help improve coverage. But it is hard to see how the scale of this deployment could be material. Such networks are often on industrial campuses, in factories, mines, ports and similar. These are not places that the public generally go to, and the total area covered will be a tiny percentage of the country. Private network operators may not want to allow public access – one of the reasons of having a private network is to control usage and privacy.

Even if the neutral host structure is limited to "hard to deploy" areas, the concept of a single shared network, with increasing integration including sharing active components, is a clear direction of travel. Most operators are already in a sharing arrangement that shares passive infrastructure, and there are national shared networks such as DNB in Malaysia[19].

Perhaps, rather than a clear role for neutral hosts, we can expect a wide range of sharing solutions, with varieties of technical integration, financial model and ownership, which help reduce the costs of deployment.

3.3 Reducing costs

In order to deliver lower costs to consumers, operators will need to reduce both their capital and operational expenditure (capex and opex). There are many categories of spending and most MNOs already exercise tight cost control, so there will be little low-hanging fruit. The easiest way to reduce costs is to spend less capex, and indeed, Omdia reports[20] that this is just what MNOs are doing, primarily as they see their wave of 5G investment come to an end. Clearly, if 6G, like 5G, resulted in a new wave of infrastructure spending, this would increase costs.

The tactics MNOs could adopt to lower opex include:

- Optimising the network for low maintenance – moving to single box solutions, retiring older generations, automating maintenance.
- Going fully on-line, removing shops and points of presence, and reducing sales and marketing efforts.
- Reducing staff in categories such as General and Admin, R&D, and deployment, and outsourcing as far as possible.
- Moving to lower cost headquarters building, and rationalising building portfolios.
- Considering increased network sharing, including active sharing.

Indeed, there is scope for new low-cost operators to emerge that outsource much of their operations and adopt a similarly disruptive model to the low-cost airlines. For example, imagine at the most extreme a mobile operator that:

---

[19] https://www.digital-nasional.com.my/
[20] https://omdia.tech.informa.com/om030639/global-telecoms-capex-declined-in-2023-as-large-telcos-completed-5g-rollouts

- Had no masts, renting space from a TowerCo.
- Had no RAN equipment, renting it from the suppliers along with a maintenance contract from them.
- Had no core network, buying it as a service from a hyper-scaler.
- Had no shops or physical presence, performing all activities on-line.
- Had no central office, using rented premises.
- Potentially, had no direct customers, selling wholesale capacity to mobile virtual network operators (MVNOs) who handled the customer relationship.

Such an operator would have as an asset their spectrum licence and other rights to be an operator and their brand. They would essentially be a project management entity, with perhaps only a few hundred staff. This would give them a dramatically lower cost base than their competitors and the ability to offer lower cost tariffs as a result. Since consumers mostly select their operator based on cost, they might win a larger share of the customer base and become increasingly profitable (at the expense of the other MNOs). To be clear, this is an extreme example to show what could be done and not a recommendation of what should be done.

One category of cost reduction that 6G proponents do agree on is the reduced use of energy – which is primarily electricity. The driver for this is more to do with the green agenda than cost reduction. This was also a goal of 5G. Whether it achieved this depends on the metrics adopted. Broadly, installing 5G on an existing cell site doubles power consumption. With 5G rolled out often to more than 50% of sites this means energy usage has increased by at least 50%. However, if measured on the power used per bit of data transmitted then this metric has reduced – networks have become more efficient. Increased overall usage neither meets a green agenda nor cuts costs hence the basic metric of overall energy usage appears more appropriate.

Implications for 6G

The implications for 6G are:

- Any capex should be avoided as much as possible, especially new hardware deployments to cell sites.
- 6G should aim for increased network automation and reduced maintenance costs.
- Network sharing, including active sharing and spectrum sharing, will likely be important tools of cost reduction.
- Approaches to reducing the absolute power consumption of networks are valuable for multiple reasons.

3.4 Achieving both simultaneously

Ensuring users are always connected will come with a cost, hence it will be challenging to improve coverage and reduce costs simultaneously. This suggests increased reliance on lower cost approaches to coverage, namely:

- HAPs and satellite, ideally owned by a third party, for rural coverage.
- National roaming for urban coverage, and for reducing congestion.
- Extensive use of Wi-Fi for indoor coverage and reducing congestion.
- Government funding for railway networks using a shared network approach.
- Local / private network provision where it is delivered and opened to others.

Many of these have minimal costs, allowing MNOs to both reduce costs and increase capacity.

Having looked at the problems that clearly currently exist, the next chapter looks at what the industry is currently suggesting for 6G.

4 Current 6G visions

4.1 Introduction

Much has already been written about what 6G should be. This material is important as it will feed into the work of the standards bodies, and the resulting standards will be designed to deliver on the vision. It is where 5G went astray – by setting out an inappropriate vision it resulted in errors such as poor choice of frequency band leading to higher cost for MNOs.

This chapter looks at what others are saying. It is not comprehensive but aims to cover the major players and research entities. It is a snapshot in time, visions may change, and others may emerge. I have quoted extensively from the various vision papers as I feel it is important to clearly and directly understand what the main players are saying.

4.2 Manufacturers

4.2.1 Ericsson

Ericsson sets out their vision[21] as follows:

> The vision for 6G is built on the desire to create a seamless reality where the digital and physical worlds as we know them today have merged. This merged reality of the future will provide new ways of meeting and interacting with other people, new possibilities to work from anywhere and new ways to experience faraway places and cultures.
>
> 6G will make it possible to move freely in the cyber-physical continuum, between the connected physical world of senses, actions and experiences, and its programmable digital representation.

---

[21] https://www.ericsson.com/en/6g

The cyber-physical continuum of 6G includes the metaverse as it is typically understood – a digital environment where avatars interact in a VR/AR world – and goes further, providing a much closer link to reality. In the cyber-physical continuum, it will be possible to project digital objects onto physical objects that are represented digitally, allowing them to seamlessly coexist as merged reality and thereby enhance the real world.

Future networks will be a fundamental component for the functioning of virtually all parts of life, society, and industries, fulfilling the communication needs of humans as well as intelligent machines. As accelerating automatization and digitalization continue to simplify people's lives, the emerging cyber-physical continuum will continuously improve efficiency and ensure the sustainable use of resources.

Countless sensors will be embedded in the physical world to send data to update the digital representation in real time. Meanwhile, functions programmed in the digital representation will be carried out by actuators in the physical world. The purpose of the 6G network platform is to provide intelligence, ever-present connectivity and full synchronization to this emerging reality.

We envision a connected and sustainable physical world that is both digitalized and programmable, where humans are supported by intelligent machines and the Internet of Senses.

Examples of important 6G use cases include e-health for all, precision health care, smart agriculture, earth monitor, digital twins, cobots and robot navigation. These use cases can be sorted into three broad use case scenarios: the Internet of Senses, connected intelligent machines, and a connected sustainable world.

In the Internet of Senses scenario, the immersive communication of 6G will deliver the full telepresence experience, removing distance as a barrier to interaction. Extended reality (XR) technology with human-grade sensory feedback requires high data rates and capacity, spatial mapping with precise positioning and sensing, and low latency end-to-end with edge cloud processing. One example will be the ubiquitous use of mixed reality in public transport, offering separate virtual experiences for each passenger, enabling them to run virtual errands, get XR guidance and have games overlaid on the physical world.

Personal immersive devices capable of precise body interaction will allow access to experiences and actions far away to better support human communication needs. At the same time, 6G networks will also add completely new communication modes with strict control over access and identities.

Examples of the types of technology that will be needed to deliver 6G use cases include zero-energy sensors and actuators; next-generation AR glasses, contact lenses and haptics; and advanced edge computing and spatial mapping technologies. From our perspective at Ericsson, we have determined that creating the 6G networks of 2030 will require major technological advancements in four key areas: limitless connectivity, trustworthy systems, cognitive networks and network compute fabric.

This is an example of what I describe as "5G, but faster and better" or more simply "5G-on-steroids". The vision is very similar to that of 5G, with key applications being the metaverse, VR, the Internet of Senses and more. Indeed, had this vision been set out for 5G it would have not been out of place. There is nothing about improved coverage and lower cost. And the "cyber-physical continuum" sounds like something out of a bad science fiction movie.

4.2.2 Huawei

The Huawei vision states[22]:

> 6G — a more advanced next-generation mobile communication system — will go far beyond just communications. It will serve as a distributed

---

[22] https://www.huawei.com/en/huaweitech/future-technologies/6g-white-paper

neural network that provides links with integrated communication, sensing, and computing capabilities to fuse the physical, biological, and cyber worlds, ushering in an era of true Intelligence of Everything.

Building upon 5G, 6G will continue the transformation from connected people and things to connected intelligence. In essence, it will bring intelligence to every person, home, and business, leading to a new horizon of innovations.

Over the next decade, in addition to continuous wireless innovations, the rise of massive artificial intelligence (AI) and the creation of massive digital twins will be the two major catalysts that fuel more technology breakthroughs. The resulting 6G will be a game-changer in terms of both the economy and society — it will lay a solid foundation for the future Intelligence of Everything.

We envision that 6G will enable the transformation from connected people and things to connected intelligence. Compared with its predecessor, 6G will offer extreme performance and realize major improvements in terms of key performance indicators (KPIs). Furthermore, it will be a key enabler in achieving the full-scale digital transformation of all vertical businesses.

More importantly, we believe that 6G will serve as a distributed neural network, providing links with integrated sensing, communication, and computing capabilities. This will fuse the physical, biological, and cyber worlds, ushering in an era where everything will truly be sensed, connected, and intelligent.

The 6G network functions as the fabric of the converged physical and cyber worlds. First, from the cyber world to the physical world — this is a typical downlink channel — the primary service will be all kinds of XR, which, enhanced by tactile communication and new human-machine interface, creates immersive experience when interacting with the digital world. Meanwhile, the continuous deep learning in the digital world serves as AI engine for the physical world, providing real-time inferences to facilitate all kinds of decision making. This imposes major challenges on the radio interface design, requiring extremely high throughput and ultra-low latency. Second, from the physical world to the cyber world — this is a typical uplink channel — the primary application is sensing and the collection of big data for machine learning (ML).

6G will be a network of sensors and ML, where data centers will become neural centers, and ML tasks will spread over the entire network, from neural center to deep neural edges (e.g., base stations or even mobile devices).

This is somewhat similar to Ericsson – 5G-on-steroids. It uses language that is often unintelligible, for example, "This will fuse the physical, biological, and cyber worlds, ushering in an era where everything will truly be sensed, connected, and intelligent." It uses the same language as was used for 5G of being a "game changer" without once noting that 5G failed to deliver on its promises. Like Ericsson it says nothing about coverage or cost.

4.2.3 Nokia

Nokia, compared to the other manufacturers, is somewhat more sensible. They state[23]:

> Over the past 30 years, cellular communications have brought unprecedented benefits to humankind. 2G and 3G unleashed the potential of mobility and connectivity. 4G gave us greater access to information and social engagement. 5G enables the enterprise and extended reality (XR) experiences. In 6G, we believe the fusion of physical, digital and human worlds will allow us to interact with the digital and physical space more intuitively, transforming the way we live and work.

---

[23] https://onestore.nokia.com/asset/214027

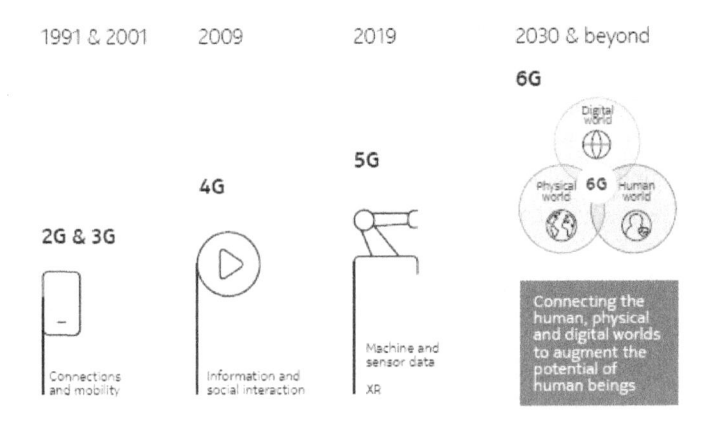

1991 & 2001	2009	2019	2030 & beyond
			6G
		5G	Digital world
	4G		Physical world — 6G — Human world
2G & 3G			
		Machine and sensor data	Connecting the human, physical and digital worlds to augment the potential of human beings
Connections and mobility	Information and social interaction	XR	

Given the 6G vision, we now describe our view of what should be supported in the first release of 6G – we call this 6G day one. A robust and extensible day one 6G baseline is of paramount importance for the initial success of 6G, as well as for its forthcoming releases to continue offering value.

Affordable and super-efficient 6G rollout. The 6G day one rollout must leverage earlier 5G investments, making it affordable with one single 6G architecture and reusing cloud-native

5G core network functions where possible with critical extensions. Initial roll-out and deployment of 6G must be highly efficient. AI will play an important role in increasing efficiency. 6G day one will be ready for rollout on existing macro-cell sites, including support for both traditional distributed RAN deployments as well as supporting cloud RAN cases. Multi-RAT Spectrum Sharing (MRSS) will ensure seamless transition between 5G and 6G for the existing bands without compromising network performance. 6G Carrier Aggregation is then used to combine coverage and capacity. Dual connectivity may be needed to achieve the same combination for non-collocated 6G sites.

Next-generation mobile broadband (MBB) services. 6G day one should support the next generation (NextGen) of affordable enhanced mobile broadband with much better performance than 5G. This is important as

> NextGen MBB is estimated to be the dominant traffic in 2030 when 6G is launched. Improved NextGen-MBB performance comes from the systems, architecture, and radio technology enablers as outlined in the previous section.
>
> *Enhanced fixed wireless access.* Excellent support for fixed wireless access (FWA) is also a priority for 6G day one. FWA has been one of the fastest-growing use cases for 5G, and the traffic volume from FWA is expected to account for up to one-third of the network traffic volume of 6G. FWA builds on NextGen-MBB and has several characteristics that can be explored for both leaner operation and higher performance. FWA customer premises equipment (CPE) has a larger form factor than a smartphone and is connected to a power outlet. A CPE is stationary, and the location of the CPE can be optimized for the network layout.

Apart from a brief mention of "the fusion of physical, digital and human worlds will allow us to interact with the digital and physical space more intuitively" this is refreshingly free of science fiction jargon. The day one focus is also somewhat realistic, recognising that 6G will need to build on 5G and noting that current broadband and FWA services will remain the most important. Spectrum and network sharing is also mentioned.

4.2.4 Samsung

Samsung entitled their vision "The next hyper-connected experience for all"[24]. They started with what they saw as the megatrends:

> In addition, today's exponential growth of advanced technologies such as artificial intelligence (AI), robotics, and automation will usher in unprecedented paradigm shifts in the wireless communication. These circumstances lead to four major megatrends advancing toward 6G: connected machines, use of AI for the wireless communication, openness of mobile communications, and increased contribution for achieving social goals

---

[24] https://www.samsung.com/global/business/networks/solutions/6g/

They noted that "It is envisaged that the number of connected devices will reach 500 billion by 2030, which is about 59 times larger than the expected world population by that time." (This is discussed further shortly.)

In terms of services, Samsung envisages:

> Representative categories of 5G services, i.e., enhanced mobile broadband (eMBB), ultra-reliable and low latency communications (URLLC), and massive machine-type communications (mMTC) will continue to improve moving towards 6G. In this section, we focus on new 6G services that will emerge due to advances in communications as well as other technologies such as sensing, imaging, displaying, and AI. Those new services will be introduced through hyper-connectivity involving humans and everything and provide ultimate multimedia experience. In the rest of this section, we highlight three key 6G services, namely, truly immersive extended reality (XR), high-fidelity mobile hologram, and digital replica

They claim that:

> However, in order to provide truly immersive AR, the density should be largely improved and it will require 0.44 gigabits per second (Gbps) throughput (with 16 million points). In addition, XR media streaming may have similar demands to 16K UHD (Ultra High Definition) quality video. For example, 16K VR requires 0.9 Gbps throughput (with compression ratio of 1/400). The current user experienced data rate of 5G is not sufficient for seamless streaming.
>
> In order to provide hologram display as a part of real-time services, extremely high data rate transmission, hundreds of times greater than current 5G system, will be essential. For example, 19.1 Gigapixel requires 1 terabits per second (Tbps) [8]. A hologram display over a mobile device (one micro meter pixel size on a 6.7 inch display, i.e., 11.1 Gigapixel) form-factor requires at least 0.58 Tbps. Moreover, support of a human-sized hologram requires a significantly large number of pixels (e.g., requiring several Tbps) [9]. The peak data rate

of 5G is 20 Gbps. 5G cannot possibly support such an extremely large volume of data as required for hologram media in real-time

They cover the cyber-physical, noting that:

> With the help of advanced sensors, AI, and communication technologies, it will be possible to replicate physical entities, including people, devices, objects, systems, and even places, in a virtual world. This digital replica of a physical entity is called a digital twin. In a 6G environment, through digital twins, users will be able to explore and monitor the reality in a virtual world, without temporal or spatial constraints. Users will be able to observe changes or detect problems remotely through the representation offered by digital twins.

> Users will be even able to go beyond observation, and actually interact with the digital twins, using VR devices or holographic displays. A digital twin could be a representation of a remotely controlled set of sensors and actuators. In this manner, a user's interaction with a digital twin can result in actions in the physical world. For example, a user could physically move within a remote site by controlling a robot in that space entirely via real- time interactions with a digital twin representation of that remote site.

Samsung are aligned with Ericsson and Huawei, with two of their trends – XR and digital replica – very much part of the cyber-physical concept. They have one difference – the importance of holograms. These were also considered to be a key element of 5G but have never proven commercially viable.

4.2.5 The focus on the cyber world

Introduction

All the manufacturers discuss the concept of digital worlds: virtual environments that may have some link to the real world. It is worth dwelling on why they have done this when all the evidence to date is that there is very little interest in such concepts.

The metaverse and VR

The virtual world was very much a part of 5G visions. Often this was called the metaverse. Researchers suggested that a virtual world experience would require data rates of 100's of Mbits/s and latencies of 10ms or less.

The virtual world concept was widely embraced by industry. In particular, Meta spent heavily on developing the metaverse – some $46bn over the last three years[25]. It has heavily subsidised VR headsets and developed software to enable users to interact in the virtual world. There was a huge spike in interest from many companies in 2022 but this quickly faded and the metaverse is now widely considered a failure, and indeed a daft idea. For example, The Nation talked about the "catastrophic failure of the metaverse"[26]. It noted:

> There was only one problem: The whole thing was bullshit. Far from being worth trillions of dollars, the Metaverse turned out to be worth absolutely bupkus. It's not even that the platform lagged behind expectations or was slow to become popular. There wasn't anyone visiting the Metaverse at all.
>
> The sheer scale of the hype inflation came to light in May. In the same article, Insider revealed that Decentraland, arguably the largest and most relevant Metaverse platform, had only 38 active daily users. The Guardian reported that one of the features designed to reward users in Meta's flagship product Horizon Worlds produced no more than $470 in revenue globally. Thirty-eight active users. Four hundred and seventy dollars. You're not reading those numbers wrong. To say that the Metaverse is dead is an understatement. It was never alive.
>
> In retrospect, that's not surprising. If you are wondering what the point of the Metaverse is—business meetings? parties? living out a kind of late-'90s Second Life fantasy but without legs?—you are not alone. In fact, no one, even Zuckerberg himself, was ever really clear what the

---

[25] https://www.nasdaq.com/articles/meta-platforms-has-spent-$46-billion-on-the-metaverse-since-2021-but-its-spending-twice-as

[26] https://www.thenation.com/article/culture/metaverse-zuckerberg-pr-hype/

whole enterprise was for except being the future of the Internet and a kind of vague hanging out.

Accessing the virtual world requires virtual reality (VR) headsets (AR and XR is discussed later). VR headset sales have languished for years at less than 10 million units per year as the data from Omdia, below, shows (quite why they expect an upturn after 2025 is unclear).

Consumer VR headset sales by category, global, 2018-2028

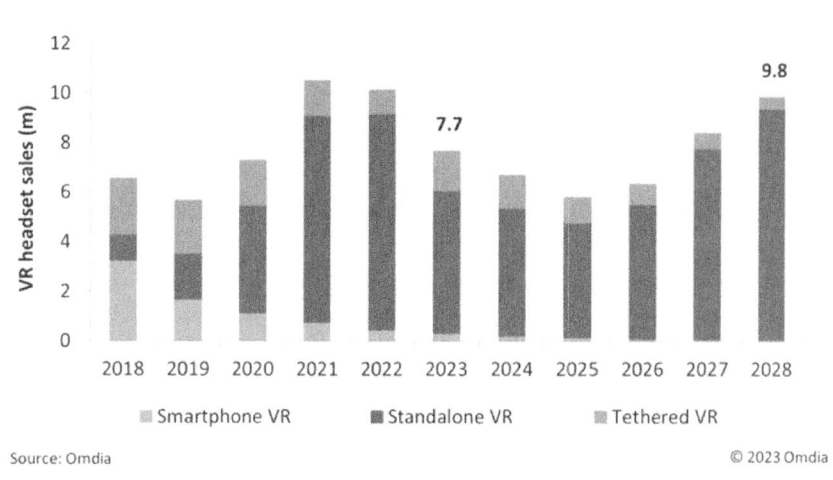

Figure 4-1 : Omdia VR headset sales and forecast

If we assume headsets have a lifetime of around three years, then this suggests something like 25m headsets in working operation globally. That is 0.3% of the population. Compare that to around 75% handset penetration. For years, virtual world proponents awaited Apple's long-rumoured VR headset in the hope that there would be the equivalent of an iPhone moment. The Vision Pro was launched in February 2024. Apart from an initial burst of interest from rich Apple fans, sales have been anaemic. It was reported[27]:

---

[27] https://www.silicon.co.uk/e-innovation/wearable/apple-vision-pro-europe-571369

> IDC estimated that Apple had yet to sell 100,000 of the devices in a single quarter since the US launch in February and faces a 75 percent drop in the domestic market in the current quarter.

More generally, VR headset sales appear to be in free-fall. IDC reported[28]:

> Global shipments of augmented reality and virtual reality (AR/VR) headsets declined 67.4% year over year in the first quarter of 2024 (1Q24).

Surely, it should be clear to all those thinking about use cases for 6G that the virtual world has been a failure despite huge investments from the world's tech giants.

Even if the virtual world had succeeded, the 5G proponents never addressed the issue that all VR headsets were Wi-Fi connected. None were cellular-connected. If there was going to be a connectivity challenge it would be met by enhanced Wi-Fi and broadband pipes, not cellular. This was blindingly obvious, yet many accepted the story that the metaverse could only emerge with 5G.

There has been a change in recent years, with more of a focus on augmented reality (AR), rather than VR. AR starts with an image of the real world and superimposes additional information. It can work either with glasses that allow the wearer to see through them and then impose a "heads-up" style display on top, or it can work with headsets that do not allow direct viewing of the real world but use external cameras to capture the real world and the display it on the screens within the headset. The Vision Pro has this latter mode of operation (as indeed, have many VR headsets, but until the Vision Pro it was rather an afterthought).

There was a brief burst of interest in wearing the Vision Pro outdoors, but this soon faded as the novelty wore off. The growth in AR is now very much in smart glasses – glasses that look like conventional sunglasses but have built in cameras and speakers. A good example is the Ray-Ban Meta Wayfarer shown below[29].

---

[28] https://www.idc.com/promo/arvr
[29] https://www.meta.com/gb/smart-glasses/shop-all/

Figure 4-2 : Ray-Ban smart glasses

These costs around $400 – just over 10% of the price of the Vision Pro. It is not strictly AR – it cannot impose images on top of vision. It can capture pictures and video, play music and react to voice commands. Glasses of this sort may well gain much larger market share than VR headsets but clearly do not need 5G, let alone 6G. Nor will they enable cyber-physical continuums.

So, the metaverse is a "catastrophic failure" and VR headsets are only owned by 0.3% of the population and sales are falling. Connectivity for virtual worlds is delivered by Wi-Fi not cellular and 5G has not been a success. Even if virtual worlds were successful and required in places where Wi-Fi was not available, 5G could deliver all that is needed. Why, then, do the manufacturers think that virtual worlds are the key driver of 6G?

I can only assume that there is some sort of collective groupthink happening amongst a close-knit community of techno-geeks. Alternatively, with no reason for a 6G vision that is 5G-on-steroids, cyber-physical mumbo-jumbo is the only way manufacturers can bamboozle the uncritical.

Digital twins and cyber-physical continuums

Another part of the same vision suggests future 6G users will mix the physical and the digital broadly through large-scale digital twins. That is, 6G stakeholders will build a model of the real world, which is kept up to date, and will use the model to influence the real world.

The broad concept of a digital twin is a simulation of something. A good example to explain the concept is an airport where this technology is in early implementation[30]. Here, the model simulates aspects such as passenger flow through the airport and predicts where there might be congestion, perhaps, for example due to a sudden gate change that causes a large flow of passengers to a different security checkpoint. The model is fed real-time data from the airport's IT system and sensors and then runs forward, predicting problems. If there are issues, then alerts can be raised, or the model itself can take action, for example sending automated messages to security staff to move to different areas.

To work the following need to be the case:

1. The modelling is tractable – it is possible to develop a good simulation.
2. The key data needed can be generated, perhaps via sensors.
3. There is an outcome that generates more value than the cost of development and deployment.

These may hold true at an airport. It is relatively easy to simulate movement of people since they follow a pre-ordained path. It is easy to count people numbers, using, for example cameras that recognise bodies. And a more efficient airport can handle more flights, gaining revenue.

It is worth noting that this example does not require humans to don VR headsets and immerse themselves in the cyber. The digital twin can simply show on a normal screen where issues will occur. Nor does this require high-speed wireless networks – cameras and similar will likely be linked via wired connections as they currently are.

Digital twins have been of much interest for around 20 years. But they have been little implemented – suggesting that the business case rarely makes sense. Where

---

[30] https://unity.com/blog/industry/how-digital-twins-are-transforming-large-scale-airports

Current 6G visions

they have been implemented it is generally in campus environments or buildings – airports, ports, oil rigs, factories, etc.

Many 6G proponents appear to be imagine that the whole world becomes a digital twin. This fails all of the necessary conditions set out above, namely:

1. The modelling is not tractable – simulating every aspect of the real world is beyond any modelling capability.
2. The key data needed cannot easily be generated – the number and type of sensors needed would be enormous.
3. There is no clear revenue model – who would pay for this hugely expensive deployment? What revenue benefits would accrue? Indeed, what would it actually be used for?

This is an example of the cellular industry not considering the wider issues of aspects such as form factors of devices, complexity of implementing sensors, challenges in simulation and more. It was an issue highlighted by SK Telecom in their assessment of why 5G failed.

Interestingly, the EU, in their briefing paper "The Path to 6G"[31] stated that:

> Advanced sensors, AI and high-speed connectivity ensuring low-latency (e.g. the time data takes to transfer across the network), would allow replication of physical entities (including people, devices, objects, systems, and even places) in a virtual world. This digital replica of a physical entity is called a digital twin. Among the various use cases, digital twins might be used in medicine to build human immune systems, or in factories to simulate the deployment of specific complex deployed assets such as jet engines and large mining trucks, to 'monitor and evaluate wear and tear and specific kinds of stress as the asset is used in the field'. One think tank stresses how 6G performance (e.g. speed and latency) will better enable digital twin technology by

---

31 https://www.europarl.europa.eu/RegData/etudes/BRIE/2024/757633/EPRS_BRI(2024)757633_EN.pdf

providing real-time monitoring and control of physical objects and environments.

The examples set out – a human immune system and a factory – are very constrained. The human immune system might require some sensors worn by patients – something that already happens. The "jet engine" example requires sensors built into the engines - as they already are - and data delivered from engines – as already happens. Neither require 6G or even 5G. The link to the "think tank" mentioned no longer works suggesting that the publication has been withdrawn.

Just how many sensors?

One of the three key pillars of 5G was "massive machine type connectivity". This was the view that huge numbers of devices such as sensors would need to be connected. This has not proven to be the case, with connected device numbers growing slowly and staying well within the capabilities of 4G's NB-IoT technology.

The charts from Transforma[32], below, are useful in gaining a sense of perspective.

---

[32] https://transformainsights.com/research/forecast/highlights

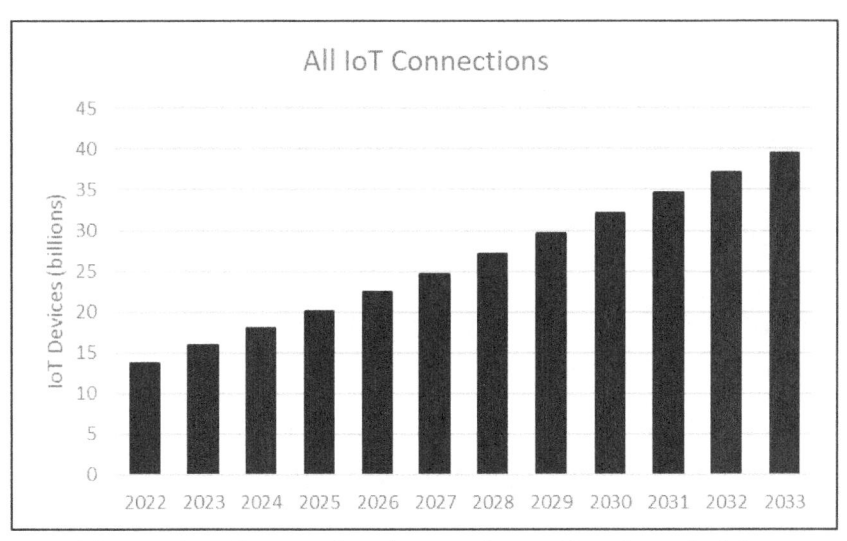

Figure 4-3 : IoT connections forecast: Source Transforma Insights

Overall IoT connections are around 18 billion – about 2 for every cellphone. Connections are forecast to double over the next ten years. But the chart below shows that few of these are on cellular networks.

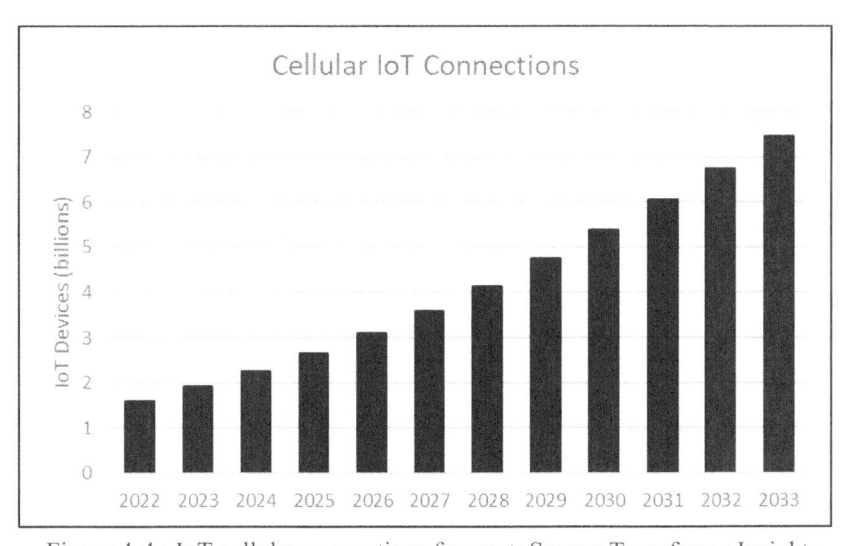

Figure 4-4 : IoT cellular connections forecast: Source Transforma Insights

Only about 2 billon devices are currently cellular-connected with numbers forecast to rise to about 7 billion by 2035. To put that into perspective, by 2035 there will be around one cellular-connected IoT device per cellphone. The vast majority – over 80% - connect via other technologies such as Wi-Fi, Bluetooth and private IoT networks running LoRa and similar. Of these cellular connected sensors many are in smart meters and similar, generating tiny amounts of data.

This is starkly at odds with data quoted by the EU[33] and Samsung[34] both of which say that there will be 500bn connected devices by 2030 and use this as an important justification for 6G. Both take this number from a Cisco report[35] although the source report is no longer available. Cisco have a poor track record of forecasting IoT connections. In 2010 they forecast that there would be 50bn connected devices by 2020, There was about 8bn. The Transforma data shows just how deeply unlikely 500bn devices is by 2030 – around 32bn is more likely of which only around 5bn will be cellular connected. This is well within the capabilities of 4G technology, let alone 5G. Rapid changes in the number of sensors is near-impossible, installing sensors is time-consuming and there is no clear reason why sensor numbers would leap.

The 500bn forecast is so far from being plausible as to suggest anyone using it really does not understand the industry they claim to be leading.

4.2.6   Manufacturers summary

It was noted earlier that the manufacturers are at the heart of the cellular generational process, driving the standards. Hence, their visions are most likely to shape the standards. Unfortunately, they have, broadly, doubled down on 5G, stating that 6G needs to be 5G-on-steroids. They appear to have entered their own world of cyber-physical continuums that bears no relation to any sort of reality or user desire. Apart from Nokia's somewhat more pragmatic stance, this

---

[33] https://www.europarl.europa.eu/RegData/etudes/BRIE/2024/757633/EPRS_BRI(2024)757633_EN.pdf

[34] https://www.samsung.com/global/business/networks/solutions/6g/

[35] https://www.cisco.com/c/dam/global/fr_fr/solutions/data-center-virtualization/big-data/solution-cisco-sas-edge-to-entreprise-iot.pdf

is deeply worrying, suggesting that 6G will be even more of a failure than 5G was.

It is alarming that these 6G visions major on applications that are so obviously failures.

4.3 Operators and operator alliances

4.3.1 NGMN

Few MNOs set out their own vision of 6G[36], tending more towards remarks at conferences and similar. Instead, they work through the Next Generation Mobile Network (NGMN) association. This has published several valuable papers which are quoted from extensively due to their important message and because they are the only material voice of the MNOs.

The first paper was "6G Drivers and Vision"[37]. This stated (with underlining added):

> We expect the fundamental goal of enabling socio-economic transformation and automated industries will naturally continue to be realized beyond this decade and beyond the 5G design goals. We expect 6G in its novelty and capabilities to meet the drivers we have outlined, to involve new advancements in pushing the envelope of performance, provide significant change in enablers, and in addition, break new frontiers (e.g. with respect to environmental impact, societal benefits, users, scenarios, players, value creation, spectrum, etc.), new business models, and potential new paradigms unknown today. We also require features of 6G to be introduced in a way that enhances trustworthiness, security and resilience.
>
> Given the forward-looking vision and design of 5G, the trends towards cognitive, autonomous, multi-access convergence, and disaggregated

---

[36] This may be partly because, as noted earlier, they tend to have hugely reduced their technical and visionary expertise, closing research and development teams and relying on manufacturers to lead the way.

[37] https://www.ngmn.org/work-programme/ngmn-6g-drivers-and-vision.html

software-based networks, and its embedded capabilities, 6G is expected to break from the historical approach of technology generations. The approach for 6G should be based on agile and fully flexible systems, with distributed intelligence including at the edge. 6G will thus be built upon the features and capabilities to be introduced with 5G, alongside novel capabilities, in order to deliver new services and value.

As indicated earlier, any new technology, nonetheless, should enable superior functionalities and capabilities, supporting radically new and differentiated services, advancing digital transformation and opening up greater market opportunities compared to those enabled by current technologies.

In its role to meet the expected goals, 6G will involve <u>enabling a seamless and ubiquitous experience, and service continuity, considering efficiency and affordability</u>. Sustainability that includes energy efficiency and adoption of green technologies and green energy, towards carbon neutrality is a key focus of NGMN, for this decade and beyond, and should be a fundamental design consideration for 6G.

This can only succeed with a holistic approach by the entire ecosystem, including global standards, ecosystem design, service footprint, metering and monitoring, and deployment strategies, among other factors. Beyond network infrastructure, this holistic approach must involve user terminal design, to foster upgrades, reusability, repairability and recycling with the goal to extend their life, <u>as well as service / applications design to optimize the amount of data to be exchanged over the networks</u>.

Some examples of attributes and design considerations are indicated below more specifically:

- Introduce new human machine interfaces that extend the user experience across multiple physical and virtual platforms, sensing, and immersive mixed realities for a variety of use

Current 6G visions

cases, including the use of large bandwidths in existing and new spectrum bands.
- <u>Advance enablement of seamless multi-access service continuity, using terrestrial and non-terrestrial networks, and provide coverage across land, sea, and sky</u>, efficiently addressing any traffic and connection density.
- Ensure cost and energy efficient delivery of heterogeneous services that have extremely diverse requirements, under the stringent constraints of energy consumption and carbon emission limits and towards achieving the goals of sustainability and carbon neutrality.
- Advance and build from design the forward-looking capabilities introduced with 5G such as disaggregation and software-based agile, cognitive and autonomous networks, to ensure the introduction of new technology plug-ins in both the network and the user terminal / interaction mechanisms, that are market driven, support innovation, and create new value opportunities.
- In support of AI by design, develop an energy and cost-efficient structure that is highly scalable, flexible, and portable, allowing abstraction and distribution of complexities, development of digital twin representation, and embedded intelligence. Identify appropriate AI-based frameworks, with the objective of supporting value creation and delivery, resource allocation optimization, and sustainable deployment and operation, among others.
- Address the future demands through the support of regulatory systems and harmonized and coordinated global standards and ecosystem, in accordance with developmental considerations outlined above, that ensures interoperability, sustainability, and massive economies of scale supporting value creation and delivery by MNOs and their partners, in a broad ecosystem.

This is notable in avoiding any mention of "cyber" and similar, while picking up on being always connected and efficiency. This was then followed with the 6G

Position Statement[38] which set out the following priorities and principles (again, underlining added).

> Operational Priorities
> 1. <u>Network simplification leading to lower operational cost whilst retaining scalability and flexible deployment models.</u>
> 2. <u>Absolute energy reduction when assessed across mobile and fixed networks to support the transition towards low carbon economies.</u>
> 3. Features (such as AI) that support automated network operations and orchestration to enable efficient, dynamic service provisioning.
> 4. <u>Proactive network management capabilities across fixed and mobile networks to predict and address issues before they impact user experience.</u>
> 5. Quantum safe infrastructure, resistant to attack by Quantum computers
>
> Guiding Principles
> 1. 6G mobile network standards must be globally harmonised.
> 2. <u>6G must not inherently trigger a hardware refresh of 5G radio access network (RAN) infrastructure.</u> The decision to refresh 5G RAN hardware for operational reasons such as end-of life, energy consumption or new capabilities must be an operator driven choice, independent of supporting 6G.
> 3. <u>6G introduction must allow certain scenarios to be realised through software-based feature upgrades of existing network elements</u> to meet 6G requirements.
> 4. 6G must not result in degraded performance for customers connected to 5G networks.
> 5. New features should be able to be deployed as and when required, without compromising existing core connectivity services such as voice.
> 6. <u>6G must address demonstrable customer needs across mobile, fixed and non-terrestrial networks.</u>

---

[38] https://www.ngmn.org/wp-content/uploads/NGMN_6G_Position_Statement.pdf

7. 6G must ensure interoperability and backward compatibility with 5G.

8. 6G must incorporate robust security measures by design to protect against emerging threats and vulnerabilities.

The message here is very clear – new hardware is to be avoided; the focus should be on reducing operational costs and on demonstrable customer needs. The next paper was 6G Requirements and Design Considerations[39]. This set out:

> When considering the communication system needs in 2030 and beyond the NGMN identified a set of 6G features that are essential. These features are not necessarily driven by new or speculative use cases but reflect essential needs when operating public networks.
>
> Digital Inclusion. <u>Everyone should be able to access digital services with a good level of service quality and in an economically accessible way. This leads to the following guidance: • 6G Design should be designed to enable coverage of sparsely populated areas in an economically viable way. • User interfaces to 6G should be simple and support intuitive interactions. • Digital inequity of 6G services should be avoided</u>
>
> Energy Efficiency. Energy efficiency, when measured by energy consumed per transmitted volume, has improved by several orders of magnitude since the introduction of mobile networks. However, the volume of data transported has increased at a greater rate which means that the total energy consumed has increased over time. This negatively impacts the environmental sustainability of networks, adds cost to operations, and threatens the ability for networks to continue to scale to meet future capacity demand. In addition, many operators have made strong commitments regarding reaching carbon neutrality and net-zero carbon emission, during the 6G-era. This leads to the following guidance: • <u>Improvements in energy efficiency must continue to be sought for 6G, and the improvements delivered should exceed the</u>

---

[39] https://www.ngmn.org/wp-content/uploads/NGMN_6G_Requirements_and_Design_Considerations.pdf

forecast growth in traffic volume to reduce overall energy consumption.
• To support this, energy consumption figures need to be comparable and interoperable between equipment suppliers and must be made available at all levels of the system to enable 6G system wide monitoring and optimisation. • Network features should be supported for low energy consumption of end user devices, and energy scavenging for IoT devices. [...]

The essential needs are particularly directed towards building networks that deliver digital inclusion, that are environmentally and economically sustainable, that reduce complexity, address traffic growth and enable new services through additional features that are complementary to existing mobile networks. At this stage, there is no decision on whether a new radio access technology or core network is required for mobile services. However, the views of NGMN indicate that for a fundamental change to take place there must be significant benefits in essential metrics such as spectrum and energy efficiency that justify the cost and complexity in technology migration. In any migration to 6G the transition should be carefully considered to ensure that 5G networks are not compromised with regards to spectral efficiency, and that the features provided by 6G provide end-user value through the addition of new features or the ability to reduce operational cost and subsequently improve affordability.

This touches on many of the points from the previous chapter – the need for being always connected, inclusion and cost reduction. Requirements are not mentioned in terms of Gbits/s or speculative services but practical needs of operators and users. Finally, the 6G Use Cases and Analysis paper[40] stated:

Whilst a variety of usage scenarios have been forecasted (in the 6G time-horizon), many could also be served over advanced 5G networks. It is challenging to identify those use cases that will be addressed specifically after 2030 and aligned with 6G. Therefore, the use cases presented here are provisional.

---

[40] https://www.ngmn.org/wp-content/uploads/220222-NGMN-6G-Use-Cases-and-Analysis-1.pdf

The operators are clearly struggling to find realistic new use cases that justify 6G.

4.3.2 Operator remarks

Individual operators have made various remarks at conferences and similar. For example, LightReading reported[41]:

> Nobody really knows what 6G will be, or even if the generational approach will still be in use eight years from now, when it would be expected to arrive. The most far-fetched ideas involve brain-computer interfaces, allowing people to control machines by thought. The most humdrum envisage more energy-efficient base stations and open interfaces. But two prominent telco executives think they already know what it will not be – and that's a new air interface.
>
> Andrea Dona, the chief network officer of Vodafone UK, and Howard Watson, the chief technology officer of BT, are increasingly confident 6G will use the same orthogonal frequency division multiplexing (OFDM) technology on which 5G is based. It is a significant development simply because much of the research for previous generations was focused on the air interface.
>
> "6G is not going to have a new connectivity layer," said Dona at a press conference this week. "It's not going to because there is nothing coming out of universities in terms of new access methodology. It is still going to be OFDM technologies, the same technologies I studied 20 or 30 years ago at the University of Padua."
>
> His remarks came days after Watson delivered the same message at the recent Mobile World Congress (MWC) tradeshow in Barcelona. "I love the fact it's the same radio standard," he told Light Reading. "So far it's looking like that's the same."

---

[41] https://www.lightreading.com/6g/6g-will-not-be-a-new-radio-standard-say-bt-and-vodafone

4.3.3 Operators summary

The views of the operators could not be more different from those of the manufacturers. The operators broadly believe that there is no need for a new air interface and that 6G should be a software upgrade concentrating on delivering greater coverage, cost reduction and flexibility. They cannot see new services that require anything beyond 5G and do not talk the same science-fiction language as the manufacturers. Later chapters will discuss how such a dichotomy could be resolved.

4.4 Research bodies

4.4.1 University of Surrey / 6GIC

The University of Surrey was a key driver of 5G visions and research through its 5G Innovation Centre (5GIC). This has now become the 6GIC. Its vision[42] is:

> 6G will enable a rich new fabric of digital services, including extending human senses and ambient data in a fusion of the virtual and physical worlds. Imagine a world where one can interact with colleagues and friends from different continents, from different cultures, without any perception of not being in the same room. Imagine extending the human experience, via digital solutions, into a realm of new sensory and tactile perceptions. Imagine interacting seamlessly with machines, and enjoying personally tailored healthcare and well-being programmes supported by extensive and yet non-intrusive sensors Imagine hyper-fine geolocation, with context-aware digital services supporting human scale activities such as physical product browsing and machine tracking. We refer to this as data teleportation. This is not the movement of atoms, as in science fiction, but the movement of information, as in science fact. As time synchronisation to microseconds and low latency levels are required, this is beyond the capabilities of 5G technology, but will be within reach with 6G. Teleportation in this form will support a range of new applications including e-health, telecare, beyond industry 4.0, and many others.

---

[42] https://www.surrey.ac.uk/sites/default/files/2020-11/6g-wireless-a-new-strategic-vision-paper.pdf

While not quite as bad as the manufacturers, it is in the same vein of cyber-physical transformation and virtual teleportation.

4.4.2 6G Flagship

The 6G Flagship at the University of Oulo in Finland, is a government-funded entity like the 6GIC, which has also sought EU funding for specific programmes. Its 6G Visions paper[43] states:

> Totally new services such as telepresence and mixed reality will be made possible by high resolution imaging and sensing, accurate positioning, wearable displays, mobile robots and drones, specialized processors, and next-generation wireless networks. Current smart phones are likely to be replaced by pervasive XR experiences with lightweight glasses delivering unprecedented resolution, frame rates, and dynamic range.
>
> 6G research should look at the problem of transmitting up to 1 Tbps per user. This is possible through the efficient utilization of the spectrum in the THz regime. Extended spectrum towards THz will enable merging communications and new applications such as 3D imaging and sensing. However, new paradigms for transceiver architecture and computing will be needed to achieve these – there are opportunities for semiconductors, optics and new materials in THz applications to mention a few.
>
> Artificial intelligence and machine learning will play a major role both in link and system-level solutions of 6G wireless networks.
>
> New access methods will be needed for truly massive machine-type communications. Modulation and duplexing schemes beyond Quadrature Amplitude Modulation (QAM) and Orthogonal Frequency Division Multiplexing (OFDM) must be developed and possibly it is

---

[43] https://oulurepo.oulu.fi/handle/10024/36430

time to start looking at analogue types of modulation at THz frequencies.

Security at all levels of future systems will be much more critical in the future and 6G needs a network with embedded trust. The strongest security protection may be achieved in the physical layer. During the 6G era it will be possible to create data markets, and thus, privacy protection is one key enabler for future services and applications.

6G is not only about moving data around – it will become a framework of services, including communication services where all user-specific computation and intelligence may move to the edge cloud. The integration of sensing, imaging and highly accurate positioning capabilities with mobility will open a myriad of new applications in 6G.

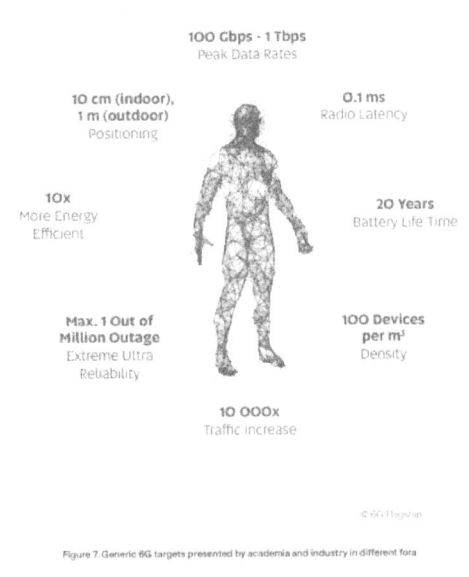

Figure 4-5: 6G Flagship vision

This feels like a repeat of 5G, with even higher targets. Broadly the same applications are mentioned as were discussed in 5G (and have failed to emerge). Unjustified targets such as 1Tbits/s are mentioned, presumably because it is much faster than 5G rather than because there is any foreseen need for it.

4.4.3 Academia summary

Academics are well-aligned with manufacturers. This may be partly because some of their funding is from manufacturers and partly because the manufacturers' agenda provides scope for substantial, interesting research. There is no discussion of any of the priorities of the operators.

4.5 Alliances and similar

4.5.1 NextG Alliance

The NextG Alliance represents interested parties in North America including operators, manufacturers and digital companies such as Google. Its vision states[44]:

> 6G is the first generation of mobile technology that places improvement to the quality of life, bridging the digital divide, and sustainability at the earliest stages of its development. 6G will deliver the next generation of technical capabilities, such as digital twins, holographic services, multi-sensory applications, ultra-high definition (HD) positioning, and network enabled robotics. These, in turn, are poised to reshape virtually every area of our economy enabling innovative solutions in areas such access to virtualized healthcare, smart agriculture, distance learning, and advanced public safety through autonomous work and transportation solutions. A coordinated, national strategy for investment in next generation solutions can simultaneously deliver 6G breakthrough developments across many industry sectors that will be determinative of the nation's economic and national security interests in the next decade and beyond.

---

[44] https://nextgalliance.org/

As is typical in the US, it flags concern over the US falling behind[45]

> Other regions of the globe have already launched ambitious partnerships between the public and private sectors to advance leadership aspirations in 6G development. China is investing heavily in bringing together government, universities, and industry to promote 6G R&D and advance China's objectives to dominate international standardization of the next generation of mobile technology. The European Commission has already committed the equivalent of nearly $1 Billion for 6G research. Japan has committed $2 Billion to encourage industrial R&D. And other countries including Korea, Finland, the U.K., Brazil, Germany, and the Netherlands have announced similar plans to advance 6G leadership in their respective countries.

Broadly, as might be expected, it sits somewhat between the manufacturers and operators, with some understanding of the need for improved coverage and sustainability, while trumpeting new services all of which appear possible over 5G.

4.5.2 6G Industry Association (6GIA)

The 6GIA represents industry bodies (often called verticals) in Europe. It is of particular importance as it works in partnership with the European Commission on the Smart Networks and Services Joint Undertaking (SNS JU) projects – the EU funding for 6G research. The EU says that "It represents the voice of European industry and research actors on 6G, bringing together operators, manufacturers, academics, small and medium-sized enterprises and ICT associations to guide the Undertaking as to where to focus research". The SNS JU is one of the largest sources of non-commercial funding for 6G research in the world with a fund of at least €1.8 billion for 2021 to 2027.

---

[45] https://nextgalliance.org/white_papers/6g-next-frontier-innovation-investment/

However, membership is not particularly balanced. Of the 43 industry players, only six are operators[46] - very much in the minority. All the major manufacturers are represented. As a result, this may not be a good source of advice for the EU on which research to fund.

The 6GIA's White Paper[47] states:

> In the coming decade, 6G will bring a new era in which billions of things, humans, and connected vehicles, robots and drones will generate Zettabytes of digital information.
>
> 6G will be dealing with more challenging applications, e.g., holographic telepresence and immersive communication, and meet far more stringent requirements.
>
> The 2030's could be remembered as the start of the age of broad use of personal mobile robotics. 6G is the mobile network generation that will help us tackle those challenges.
>
> 6G will likely be a self-contained ecosystem of artificial intelligence. It will progressively evolve from being human-centric to being both human- and machine-centric. 6G will bring a near-instant and unrestricted complete wireless connectivity. A new landscape will also emerge for the enterprises, as a result of the convergence that 6G will allow in the fields of connectivity, robotics, cloud and secure and trustworthy commerce. This will radically reshape the way enterprises operate. In short, 6G will be one of the basic foundations of human societies of the future.
>
> To enable a sustainable progress for society, in line with the United Nations Sustainable Development Goals, it is crucial that 6G addresses effectively pressing societal needs, while delivering new functionalities. This (r)evolution must be in line with Europe's primary

---

[46] See https://6g-ia.eu/industry/ . Namely DT, Orange, Telefonica, TIM, TurkCell and Vodafone.
[47] https://5g-ppp.eu/wp-content/uploads/2021/06/WhitePaper-6G-Europe.pdf

> societal values, in terms of e.g., privacy, security, transparency, and inclusiveness.
>
> Digital technologies are also becoming a critical and essential means of ensuring countries' sovereignty. The development of Europe-based 6G infrastructures and solutions is one of the keys to secure European sovereignty in critical technologies and systems. [...]
>
> To ensure that 6G can be inclusive for all people across the world, it needs to be affordable and scalable, with a great coverage everywhere. Key features of 6G will include intelligent connected management and control functions, programmability, integrated sensing and communication, reduction of energy footprint, trustworthy infrastructure, scalability, and affordability

As befits a body representing industry, the focus is unsurprisingly on connecting things and robots. The vision feels rather random, flitting across topic areas, covering technology, sovereignty and then inclusivity. In the past, industry bodies have had little input into the generational process but its input to EU funding drives a research agenda that, historically, has inexorably led to technologies being selected for generations, and then problems trying to be found for the solutions developed.

The link between 6GIA and the EU can be seen in the EU's own 6G vision statement[48] which states:

> The next generation mobile system, 6G, is described as a distributed intelligent network (underpinned by AI and machine learning), which creates 'interactions between the physical world, digital world, and biological (human) world, especially emphasising the real-time integration of cyber and physical spaces'. It is considered that 6G will improve applications of previous mobile generations and introduce new ones, such as truly immersive extended reality (XR), high-fidelity mobile hologram and digital twins of real-world objects

---

[48] https://www.europarl.europa.eu/thinktank/en/document/EPRS_BRI(2024)757633

> However, with the promise of unprecedented capabilities comes a host of challenges. Critical aspects that demand attention in the development of 6G networks are privacy and cybersecurity. As 6G aims to push the boundaries of connectivity, enabling innovations such as holographic communication, seamless extended reality, and the integration of artificial intelligence (AI) on a massive scale, the potential risks to privacy and cybersecurity are magnified (e.g. mass data collection). Another critical aspect is its environmental footprint. While 6G aims for energy efficiency, the increasing demand for data and connectivity may still pose challenges related to energy consumption. Balancing technological progress with environmental considerations remains a key objective for the development of 6G

This helps explain why much of the academic research is aligned with the manufacturers' interests and not the operators' ones.

4.5.3 ITU

The ITU, in principle, is of great importance. Its recommendations are seen by some as near-gospel. It has great stature as the world's global telecoms regulatory body. ITU graphics and targets were used to set the 5G standards goals. It terms 6G "IMT-2030" and sets out its views as follows[49]:

> Applications and services enabled by IMT-2030 are expected to connect humans, machines and various other things together. With the advances in human-machine interfaces, interactive and high-resolution video systems such as extended reality (XR) displays, haptic sensors and actuators, and/or multi-sensory (auditory, visual, haptic or gesture) interfaces, IMT-2030 is expected to offer humans immersive experiences that are virtually generated or happening remotely. On the other hand, machines are envisaged to be intelligent, autonomous, responsive, and precise due to advances in machine perception,

---

[49] https://www.itu.int/dms_pubrec/itu-r/rec/m/R-REC-M.2160-0-202311-I%21%21PDF-E.pdf

machine interactions, to the extent practicable, and demonstrated actionable management of artificial intelligence (AI).

In the physical world, humans and machines are expected to continuously interact with each other, working with a digital world that extends the real world by using a large number of advanced sensors and AI. Such a digital world not only replicates but also affects the real world by providing virtual experiences to humans, and computation and control to machines.

IMT-2030 is expected to integrate sensing and AI-related capabilities into communication and serve as a fundamental infrastructure to enable new user and application trends. From these trends, it is expected that IMT-2030 provides a wide range of use cases while continuing to provide, inter alia, direct voice support as an essential communication. Furthermore, IMT-2030 technology is expected to drive the next wave of digital economic growth, as well as sustainable far-reaching societal changes, digital equality and universal connectivity. IMT-2030 is expected to further enhance security and resilience.

Usage scenarios of IMT-2030
IMT-2030 is expected to expand and support various user, application and technology trends while providing prospects towards a sustainable digital transformation.

IMT-2030 is expected to be built on overarching aspects which act as design principles commonly applicable to all usage scenarios. <u>These distinguishing design principles of the IMT-2030 are including, but are not limited to, sustainability, security and resilience, connecting the unconnected for providing universal and affordable access to all users independent of the location, and ubiquitous intelligence for improving overall system performance.</u>

Usage scenarios of IMT-2030 are envisaged to expand on those of IMT-2020 (i.e. enhanced mobile broadband (eMBB), ultra-reliable low latency communications (URLLC), and massive machine type communications (mMTC) introduced in Recommendation ITU-R M.2083) into broader use requiring evolved and new capabilities. In addition to expanded IMT-2020 usage scenarios, IMT-2030 is envisaged to enable new usage scenarios arising from capabilities, such as artificial intelligence and sensing, which previous generations of IMT were not designed to support. The usage scenarios of IMT-2030 include:

Immersive Communication. This usage scenario extends the enhanced Mobile Broadband (eMBB) of IMT-2020 and covers use cases which provide a rich and interactive video (immersive) experience to users, including the interactions with machine interfaces. This usage scenario covers a range of environments, including hotspots, urban and rural, which arise with additional and new requirements compared with those of eMBB from IMT-2020. Typical use cases include communication for immersive XR, remote multi-sensory telepresence, and holographic communications. [..].

Hyper Reliable and Low-Latency Communication. This usage scenario extends the Ultra-Reliable and Low-Latency Communication (URLLC) of IMT-2020 and covers specialized use cases that are expected to have more stringent requirements on reliability and latency. This is typically for time-synchronized operations, where failure to meet these requirements could lead to severe consequences for the applications. Typical use cases include communications in an industrial environment for full automation, control and operation. [..].

Massive Communication. This usage scenario extends massive Machine Type Communication (mMTC) of IMT-2020 and involves connection of massive number of devices or sensors for a wide range of use cases and applications. Typical use cases include expanded and new applications in smart cities, transportation, logistics, health, energy, environmental monitoring, agriculture, and many other areas such as those requiring a variety of Internet of Things (IoT) devices without battery or with long-life batteries. [...]

Ubiquitous Connectivity. This usage scenario is intended to enhance connectivity with the aim to bridge the digital divide. Connectivity could be enhanced, inter alia, through interworking with other systems. One focus of this usage scenario is to address presently uncovered or scarcely covered areas, particularly rural, remote and sparsely populated areas. Typical use cases include, but not limited to, IoT and mobile broadband communication.

Artificial Intelligence and Communication. This usage scenario would support distributed computing and AI applications. Typical use cases include IMT-2030 assisted automated driving, autonomous collaboration between devices for medical assistance applications, offloading of heavy computation operations across devices and networks, creation of and prediction with digital twins, and others. [..].

Integrated Sensing and Communication. This usage scenario facilitates new applications and services that require sensing capabilities. It makes use of IMT-2030 to offer wide area multi-dimensional sensing that provides spatial information about unconnected objects as well as connected devices and their movements and surroundings. Typical use cases include IMT-2030 assisted navigation, activity detection and movement tracking (e.g. posture/gesture recognition, fall detection, vehicle/pedestrian detection), environmental monitoring (e.g. rain/pollution detection), and provision of sensing data/information on surroundings for AI, XR and digital twin applications. [..]

This feels somewhat like a cut-and-paste of the 5G objectives with the wording "extends 5G" added to most usage scenarios. Applications are near-identical to 5G. Building on the wide adoption of their "triangle graphic" for 5G, the ITU produced an equivalent for 6G.

Current 6G visions

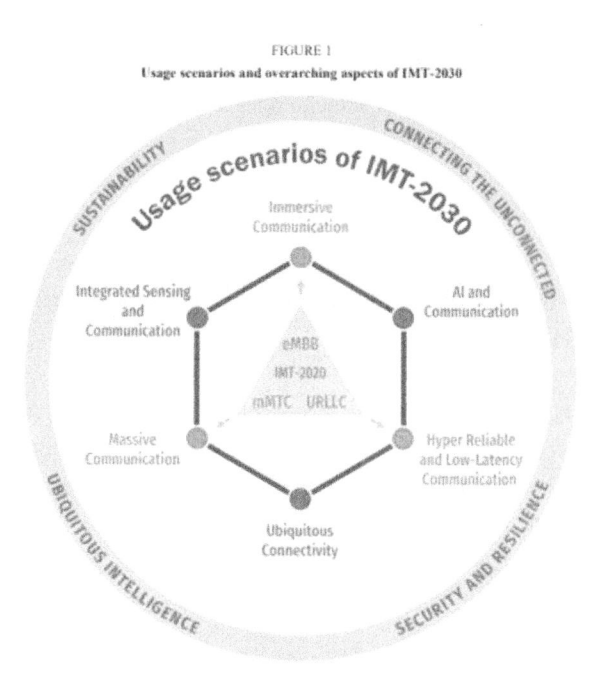

FIGURE 1
Usage scenarios and overarching aspects of IMT-2030

[Hexagonal diagram showing Usage scenarios of IMT-2030: Immersive Communication, AI and Communication, Hyper Reliable and Low-Latency Communication, Ubiquitous Connectivity, Massive Communication, Integrated Sensing and Communication; with central triangle containing eMBB, IMT-2020, mMTC, URLLC; surrounded by overarching aspects: SUSTAINABILITY, CONNECTING THE UNCONNECTED, SECURITY AND RESILIENCE, UBIQUITOUS INTELLIGENCE]

Figure 4-6 : ITU 6G graphic

This keeps the 5G triangle and aims to show how it is extended (the arrows from the points of the triangle) and with new capabilities added to the hexagon. Of these:

- Ubiquitous connectivity is not new and was part of the eMBB vision (but is welcome).
- AI and communications is unclear.
- Integrated sensing and communications is a new concept for 6G and is discussed further below.

It feels rather like the ITU could not really find a need for 6G but felt it had to produce something. With its vagueness, this ITU vision could be read by different entities in different ways but seems unlikely to be a significant influence.

4.5.4 Summary for associations and similar

The message from these bodies is weak, likely diluted by the need to achieve consensus among a diverse set of stakeholders and authors. It is hard to see how it could have any material impact.

4.6 Overall summary of visions

There are two completely separate camps – the manufacturers and academics arguing for 5G-on-steroids, and the operators pleading for a software only cost-reducing update.

This was not the case for 5G. While the operators may have had their concerns, they publicly toed the line on how 5G would change the world. How this impending conflict between the manufacturers and the operators will play out is the subject of a later chapter.

Some key take-aways from the manufacturers' visions are:

1. The desire for ever-faster speeds will drive the need for more spectrum at ever-higher frequencies.
2. There is an aim to deploy a new air interface delivering greater spectrum efficiency.
3. The broad thrust of the expected need is around a metaverse-like vision of a cyber world.
4. AI gets many mentions but with little clarity as to what it can realistically achieve.

The operator visions are generally aligned with the views developed in Chapter 3 that no new features are needed, and key is reducing cost while ensuring users are always connected at sufficient quality. The next chapter looks at some of the implications of the manufacturers' vision.

5 Assessing what the visions call for

5.1 *The visions are not clear on what is needed*

There are two sets of visions – from manufacturers (and others) and from operators. The visions from the operators are clear on what is needed – broadly nothing. The operators do not want any new hardware, believe that 5G can handle any conceivable use case, and mostly ask for lower cost and greater coverage. Chapter 3 set out how these can be achieved.

The visions from manufacturers, academics and similar - which I've termed "5G-on-steroids" - are less clear on what is needed. While there is much discussion on the cyber-physical continuum there is little on what data rates, latency and capacity is required to deliver them. Given that metaverse-style applications including immersive telepresence were a key element of 5G design, it is unclear whether 5G can deliver what is needed or not (operators clearly think it can).

The only manufacturer to clearly set out numerical performance requirements was Samsung. They produced[50] the figure below showing, for example a 50-fold increase in peak data rates, a 10-fold increase in user experienced rates and a 10x reduction in latency all while being twice as energy efficient and 100x as reliable.

Based on the evidence available, this chapter tries to piece together what the manufacturers' 6G might look like.

---

[50] https://www.samsung.com/global/business/networks/solutions/6g/

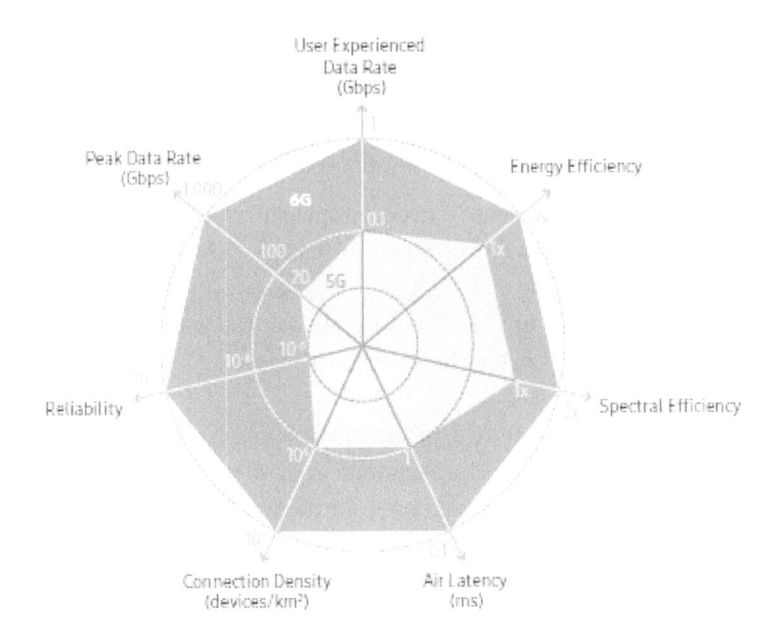

Figure 5-1 – Samsung's view of metrics for 6G

5.2 Calls for more spectrum abound

All the 5G-on-steroids visions call for more spectrum, presumably to deliver higher data rates and increased capacity. Spectrum discussions are in their early stages with the band 7-14GHz (approximately) being of much interest for the bulk of 6G applications, and the THz band (roughly above 100GHz) being seen as one of the distinguishing features of 6G.

Since the aim of this vision of 6G is higher data rates – the 6G Flagship and Samsung suggest 1Tbits/s as a target – then larger bandwidths will be needed. 5G uses bandwidths of around 100MHz to deliver its higher data rates, and with 3-4 operators in most markets this implies a total bandwidth of around 400MHz. If 6G is to be materially faster – say at least 10x user experience improvements as suggested by Samsung - then it will need about 10x the bandwidth since there is little that can be gained in spectrum efficiency. This implies a spectrum allocation of around 4GHz.

Assessing what the visions call for

Such a large bandwidth is highly unlikely to be available in the 7-14GHz band – it would take over half of that band. Even finding a few hundred MHz is proving difficult due to the extent that the band is already used. In the best case perhaps 2x the 5G allocation might become available, but a 2x improvement in data rates is not particularly material (in that it is unlikely that there will be applications that cannot function with 5G rates of around 100Mbits/s but work well with 200Mbits/s).

A bandwidth of 4GHz or more is viable in the mmWave bands (above around 30GHz) and easily found in the THz bands. Samsung focusses on THz systems and sets out the research it considers to be important to make them work well.

But all of this is deeply flawed. 5G, operating at 3.5GHz, has far weaker coverage than 4G in most countries, and the cost of improving this is prohibitive. 6G operating in 7-14GHz would likely only have around half to a quarter of the range and so likely less than half the coverage of 5G.

The use of mmWave and THz bands is even worse. 5G includes the bands at 26-30GHz for high-capacity solutions. Some operators, such as those in South Korea, have tried to use these bands and found them unworkable, handing back the spectrum in some cases. There have been very limited deployments elsewhere, none of which has been any sort of a success. The consensus view is that 5G mmWave deployments are highly unlikely. Attempting to use these bands and much higher frequencies for 6G flies in the face of experience.

6G proponents might point to their idea of using reflective intelligent surfaces (RIS) to overcome the propagation limitations of higher frequency bands. RIS are large panels, attached to walls and similar, that act as mirrors to radio waves, ideally in an active manner, changing their focus as users move. To really make a difference and extend mmWave cells significantly these would need to be plastered across many walls. The cost would be high since they require power and communications – indeed they start to look like a base station in terms of deployment issues. Building owners would expect significant rent and would not want a RIS on a glass façade. They also tend to block signals at other frequencies, or worse create intermodulations and unwanted reflections that can cause interference to other users. And all they would achieve is to fill in a few holes in already extremely patchy coverage.

If it is to deliver on the promise of improved coverage and inclusivity, then 6G needs lower frequencies than 5G. In lower bands there is less bandwidth, so this cannot provide higher data rates.

5.3 AI-native is trendy but vague

Most visions of 6G now call for it to be "AI native". This is ill-defined but the premise appears to be that it is difficult to apply AI to 5G because of the way it has been designed, and that a different approach would make the application of AI easier. The reason for these calls is clearly the huge interest that has erupted in AI – predominantly large language models – coupled with an assumption that AI will somehow breath its magic on mobile networks, delivering better, faster, more spectrally efficient solutions.

There is much doubt as to whether AI is well-suited to handling the multi-dimensional complexity of mobile networks. Even if it did work, if there is little need for more capacity then the benefits of network optimisation are minimal. There are some sensible potential uses of AI, these include automated customer care (intelligent chatbots), better fraud detection, better security and predictive maintenance. All of these can be done now, without needing a new generation. Samsung set out their thoughts on AI[51] but only came up with:

- Improve performance of handover operation taking into account network deployments and geographical environments.
- Optimize network planning involving base station (BS) location determination.
- Reduce network energy consumption.
- Predict, detect, and enable self-healing of network anomalies.

All of these are done to some extent today, would likely make a minor improvement to the network and do not merit a new generation.

---

[51] https://www.samsung.com/global/business/networks/solutions/6g/

Ideas as to how to apply AI to those aspects of networks that are affected by standards (and hence new generations) are thin on the ground at present. There are vague suggestions that MIMO beamforming antenna patterns and tracking could be improved and that the overhead of sending channel state information might be reduced. There does not appear to be any serious thinking that entire networks could be automatically optimised by AI in a manner that delivered material gains (and did not risk destabilising the network through hallucinations). It is also worth noting that intelligent optimisation solutions have been used for many years by operators, albeit based more on machine learning than artificial intelligence. These solutions may already have realised most of the possible gains.

Even if all of that is set aside, and it is assumed that there are substantial, worthwhile benefits that AI can deliver, making networks "AI native" is unlikely to require much change. To give AI the greatest chance of working as much data needs to be gathered from the network as possible, and the levers of control centralised, allowing one AI engine to control them all (otherwise there is a risk of multiple AI engines working against each other). This would suggest an architecture somewhat akin to the Open-RAN (O-RAN) concept of having a controller (the RAN Intelligent Controller - RIC) which can set RAN parameters and centralising this as is common in cloud-RAN (C-RAN) architectures. This feels like a minor change that could readily be accomplished with a 5G revision.

5.4 *Sensing is new but ill-conceived*

A new concept, mentioned by many, is the idea that 6G will add sensing to its capabilities. The 6G Flagship says[52]:

> Following the trend initiated in the 5G new radio (NR) systems, sixth generation (6G) will continue to develop towards even higher frequency ranges, wider bandwidths, and massive antenna arrays. In turn, this will enable sensing solutions with very fine range, Doppler and angular resolutions, as well as localization to cm-level degree of accuracy. Moreover, new materials, device types, and reconfigurable surfaces will allow network operators to reshape and control the

---

[52] https://www.6gflagship.com/6g-white-paper-on-localization-and-sensing/

electromagnetic response of the environment. At the same time, machine learning and artificial intelligence will leverage the unprecedented availability of data and computing resources to tackle the biggest and hardest problems in wireless communication systems.

6G systems will be truly intelligent wireless systems that will not only provide ubiquitous communication but also empower high accuracy localization and high-resolution sensing services. They will become the catalyst for this revolution by bringing about a unique new set of features and service capabilities, where localization and sensing will coexist with communication, continuously sharing the available resources in time, frequency and space. Applications such as THz imaging and spectroscopy have the potential to provide continuous, real-time physiological information via dynamic, non-invasive, contactless measurements for future digital health technologies. 6G simultaneous localization and mapping (SLAM) methods will not only enable advanced cross reality (XR) applications but also enhance the navigation of autonomous objects such as vehicles and drones. In convergent 6G radar and communication systems, both passive and active radars will simultaneously use and share information, to provide a rich and accurate virtual image of the environment. In 6G, intelligent context-aware networks will be capable of exploiting localization and sensing information to optimize deployment, operation, and energy usage with no or limited human intervention.

A first point of note is that sensing may require frequency bands at mmWave or THz (because the accuracy of resolution is related to the wavelength used). An earlier part of this chapter dealt with the issues associated with these bands.

It is also relevant that sensing already exists. Wi-Fi routers use sensing, in the form of measuring the reflections of their carrier signals, to be able to detect movement in a building and hence work as intruder alarms or provide monitoring of elderly occupants. Some phones have lidar – every pro model iPhone since the iPhone 12 for example has had lidar - delivering accuracy of around 3-5mm using frequency bands near to visible light (and hence much higher than even the THz bands). Any sensing delivered by 6G will be incremental to these capabilities.

Finally, the applications seem very weak. For example:

- Many health sensors are already delivered using existing technology such as smart watches.
- Autonomous objects already navigate with their own lidar or visual solutions and implementing a full 6G chipset would seem significant over-kill.
- The need for 6G "radar" is unclear and lidar scanning of the environment is likely to be better.

The real challenge with sensing is the need for higher frequencies. When coupled with a very vague concept of what applications might need sensing, this is another vision that has not been thought through.

5.5   It adds up to more expense than 5G for less inclusivity

As mentioned at the start of the chapter, the way the 5G-on-steroids visions map to network requirements is very unclear. Indeed, it is striking that there is so little written about this. But most aspects of the vision including higher speeds, more capacity, lower latency and sensing, all point to using higher frequency bands than 5G, in some cases much higher. This implies many more cells which are expensive, and because capex is limited also implies very limited coverage. 5G is far less available than 4G to most subscribers in most countries. While this may improve a little, it is unlikely to change materially now that many operators are scaling back further 5G deployment plans. 6G will be even less available.

In "The 5G Myth", published in 2016, I showed how the visions of 5G in vogue at the time not be economically viable or widely available. These issues were clearly predictable. Yet 6G as put forward by manufacturers and others appears not to be able to foresee the serious flaws in their concepts. And that assumes that there would be demand for 6G capabilities where available whereas all the experience with the metaverse and VR over recent years has shown that there is no significant demand for such concepts.

Doubts are creeping in. Nokia and Ericsson seem increasingly keen to disassociate themselves from using THz spectrum, albeit they often say that it will not be needed in initial deployments rather than ruling it out completely. Yet much of the research work has been around THz and associated concepts such as RIS, and Samsung and Huawei still appear to see THz as a critical element of 6G. Quite what is needed in terms of features for 6G seems unclear and in recent a recent blog[53] Ericsson suggested that perhaps a new air interface was not needed, but spectrum sharing and ever-larger MIMO antennas would be key features. They then spent much of the blog covering relatively low-level changes to the protocols which could probably be handled by a normal 5G release.

As the operators observed, most visions, however unlikely, can be met with 5G. Building on the visions to find a compelling story as to what the key features and architecture of 6G should be is proving very hard.

---

[53] https://www.ericsson.com/en/blog/2024/5/future-6g-radio-access-network-design-choices

6 A better 6G vision

6.1 Introduction

Chapter 3 set out the current problems with communications networks, based on user surveys and discussed various ways to solve them. Chapter 4 then looked at the views expressed by stakeholders and showed that the views of the MNOs tended to align with those of Chapter 3, whereas the manufacturers and academics proposed 5G-on-steroids. Chapter 5 demonstrated the impracticality of the manufacturers' visions.

This chapter builds on all this material to set out a clear, simple vision for 6G that would deliver what users actually want.

6.2 What is not needed

Because each new generation to date has always been faster and had increased capability, it is worth starting by noting that these are not needed.

There are very few situations where users will notice any benefits from data rates above 5Mbits/s since external factors, such as Internet protocols and far-end servers will become the constraining factor. With 5G delivering data rates well above 100Mbits/s, users already have vastly more than they need. It is also very hard to envisage any non-user (eg machine) applications where speeds higher than 100Mbits/s are required. Hence, there is no need for 6G to be any faster than 5G.

As set out in "The End of Telecoms History", the rate of growth of data usage rates on fixed and mobile networks is slowing. It is now around 20%/year on most mobile networks and the trajectory is to fall by around 5%/year. By around 2028 there will be minimal growth in data usage, absent any widely adopted new applications. "The End of Telecoms History" showed why no likely applications would materially change this. There may be some residual need for capacity enhancements in 5G networks, but this will occur (and hopefully be resolved) long before 6G emerges. Chapter 3 set out some ways that capacity challenges could be met within existing standards.

If more capacity is not needed, then neither is more spectrum. In the past, finding, clearing and auctioning spectrum has been a core part of each new generation. No longer. This will be a relief for MNOs who have often had to pay many £bns in auction fees.

More generally, 5G could be regarded as "better, faster" in every way – higher data rates, lower latencies, higher spectrum efficiencies, increased device capacity, etc. None of this is needed in 6G. Arguably it was not needed in 5G. As noted by the NGMN, it is very hard to envisage any applications that cannot be met with the capabilities of 5G networks. If "better, faster" is not needed then neither is a new air interface, nor any new hardware in the base stations.

6.3 What 6G needs to deliver

Drawing on the previous chapters, 6G needs to enable:

- Always connected and high quality coverage, including in rural areas, urban not-spots, indoors, and in trains.
- Highly reliable networks.
- Lower consumer costs, likely achieved through lower operator capex and opex including reduced energy consumption.

The key mechanism to achieving most of this is multi-network operation – using HAPs and D2D satellite in rural areas, national roaming in urban areas, Wi-Fi indoors, community and private networks where open, and network sharing of infrastructure.

6.4 Achieving multi-network operation

Multi-network operation – using multiple cellular networks, Wi-Fi and satellite - works today, to some degree. It is mostly achieved by the handset, which can decide to use a satellite (the iPhone emergency service) or to attach to Wi-Fi. Handsets can decide to roam to other networks, but typically this is only enabled when outside the home country. Mechanisms like Wi-Fi calling, can be standardised in cellular and implemented, albeit it took many years and has many issues. Voice calls can be made using the native cellular voice provision or over-the-top using calling apps such as WhatsApp. If handset-led multi-network operation continued to be the way to select between different networks it would be possible to deliver the kind of multi-network operation discussed in this book, but it would be sub-optimal.

Broadly, there are three ways that multi-network operation could be delivered:

1. Decentralised to the device, as today, albeit the device is given rules set by a mix of MNOs and device manufacturers.
2. Making one of the networks the priority network and requiring other networks to inter-work with the priority network. This is the current approach for native voice which is anchored back to the cellular network even when provided over Wi-Fi.
3. Implementing a new centralised control system sitting above all of the access networks.

The objectives of multi-network operation are:

1. Ensure the device is connected to the best network for its current location and user requirements.
2. Ensure that the device is contactable – that it can receive incoming calls and messages regardless of its form of connectivity.
3. Ideally delivering device independence, so that a user can use any device – phone, laptop, etc – and can still have all of their messaging available.

The second point – contactability – is worth explaining further. It is where phones differ from laptops, cellular from Wi-Fi. You can only receive an email when logged in – eg on a laptop when the email client (eg Outlook) is running. This client then periodically asks the server whether there are messages for it. But you can receive a phone call whenever your phone is turned on. To call you someone just needs your number and nothing more. Contactability is the reason why there is voice-over-Wi-Fi – otherwise if only connected by Wi-Fi and not in cellular coverage, it would not be possible to receive incoming calls made to the phone number. Contactability remains critical, for example to be able to get an SMS to verify identity (using a one-time-passcode).

It is possible to receive incoming calls on messaging apps like WhatsApp. But this only works if the caller knows you are on WhatsApp and is pre-registered as a contact. So instead of just giving, eg a car rental company, your mobile number, you would need to set them up as a WhatsApp contact and they need to do the same. It could be made to work, but it would be much less convenient than today's approach, and likely unworkable for organisations such as plumbers who would want calls from potential clients without needed to set up WhatsApp contact arrangements with them first.

Contactability currently works through a unique phone number which contains routing information. From the number, any network can work out what cellular (or fixed) network the number "belongs to" and route a call accordingly. The destination network maintains a register (the HLR/VLR combination) which knows in which set of cells that subscriber is currently located and can "page" the device. The link between number and network is hard-wired – a number always points to a particular network. Hence, this approach can only work with multi-network operation if one network is the priority network and knows how to deal with incoming calls.

A centralised approach keeps track of what devices a user is currently using, and how they are currently connected. It also moves their devices between networks as needed (although devices might also do this autonomously). It routes incoming calls and messages appropriately. It might also handle authentication and national legal requirements such as lawful intercept and emergency calling. To be clear, it does not replace the core in a cellular network but will interact with it, for example, to move devices across networks.

The ability of the three architectural approaches above to deliver against the three objectives is summarised below.

A better 6G vision

	Best connected	Contactable	Device independence
Decentralised	**Poor**. Devices have limited knowledge of what would be best	**Poor**. Unless approaches such as voice-over-Wi-Fi are implemented once off cellular the device is not contactable.	**Good**. Different devices can be used on different networks, but only a cellphone on cellular.
Priority network	**Poor**. One network has little knowledge of other networks	**Excellent**. The priority network handles all contact requests.	**Poor**. Cellphones have to be used.
Centralised control	**Excellent**. A centralised network can have a clear view across all networks.	**Excellent**. The centralised core handles and routes all incoming calls and messages.	**Excellent**. Any device can be used on any network (subject to network policies).

Table 6-1 – Options versus requirements

The table shows strong advantages for centralised control. To be clear, all other approaches can work (and do currently work) but have disadvantages, which is why we still have a world without seamless roaming across Wi-Fi, without national roaming and with future challenges in integrating satellite connectivity.

A new standard allows a re-think as to how to structure our telecoms networks. 6G should move to a centralised control approach.

6.5 The multi-network coordinator

Overview

As shown above, the optimal way to deliver true multi-network operation is to have a coordination function that sits outside of any individual network so that messaging can be routed and devices moved to the optimal network. WhatsApp might be seen as an early manifestation of this, delivering services regardless of the mode of connectivity to multiple devices simultaneously (eg a phone and a laptop), keeping track of the user's mode of connectivity and using common authentication across all networks. But it does not control what network(s) devices opt to connect to and only allows incoming calls and messages from pre-registered contacts who are also using WhatsApp.

The fixed network is configured to work with a coordinating function, with little in the way of a core network, simply acting as a bit pipe to route traffic to peering points and destination networks. But the mobile network is the opposite, with core networks of ever-growing complexity delivering ever-more services including trying to centralise voice control even when delivered over Wi-Fi.

The level of functionality of the multi-network coordinator is open for debate. As a minimum it must:

- Know how each registered user is currently connected, both network and device, so as to route incoming calls and messages.
- Be able to route outgoing calls and messages to other multi-network coordinators.

It would be highly beneficial if it:

- Moved users across networks so they were optimally connected.
- Managed user authentication so that individual networks did not need to do this themselves.
- Managed associated legal requirements since without authentication it will be difficult for individual networks to provide these.

It could, optionally also:

- Manage all billing relationships, providing a one-stop shop for users.
- Delivered API services to applications that could work across all underlying networks.

It might be asked why not just use WhatsApp? This provides multi-network functionality across multiple devices and routes incoming messaging accordingly. But as noted above, there are issues:

1. The need to set up a contact before being able to call or receive calls from them.
2. The need for everyone to use WhatsApp, since it is not compatible with other messaging apps, creating a dangerous global monopoly of messaging provision.

That is not to say that WhatsApp cannot be part of the solution, but that it should be possible for each user to select their preferred messaging platform and still be able to contact others on different platforms, without even needing to know what platform they are using. This is another function of the multi-network coordinator, knowing what messaging platform a user is using at any time and routing calls to that platform.

There are many functions that need to remain within individual networks. For example, the mobile network needs to know a user's location within the network so that it can route incoming messages. It may need to gather data on usage for billing (although fixed networks forgo this by offering unlimited data usage and removing voice calls). It needs to manage user mobility by handing off users between cells. It needs to provide operations and maintenance capabilities by monitoring network functionality and raising alarms. But other functions including authentication, service provision, APIs and user data, could be managed externally. Indeed, it already is for roaming users, where they are authenticated by their home network rather than the network they are connected to. Messaging, including voice and text-style messages should not be delivered by any individual network but run either by the multi-network coordinator or by an external messaging provider.

The Figure below shows schematically how this might look:

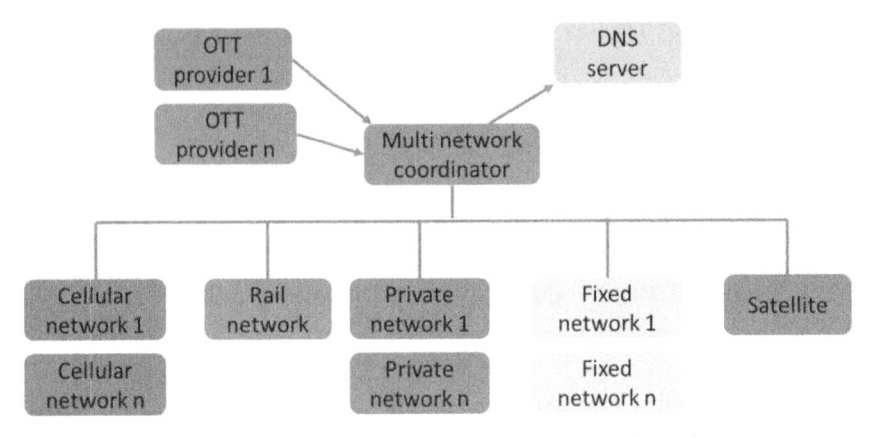

Figure 6-1 – A multi network coordinator schematic

In this figure the multi network coordinator sits above multiple networks. It has access to all relevant OTT providers so each subscriber can select their preferred OTT provider, and it communicates with a DNS server (or similar) to discover which multi network coordinator any given identity is currently associated with.

Benefits

Moving device control to a multi network coordinator would:

- Make it much simpler to integrate private and neutral host networks into the broader cellular network infrastructure.
- Make integration with satellite networks simpler since the multi network coordinator would be aware of whether a device was currently registered on a satellite or terrestrial network and route traffic accordingly.
- Make Wi-Fi integration simpler since the same authentication credentials could be used for both Wi-Fi access and cellular access (without needing the agreement of the MNO or Wi-Fi owner) and there would be visibility of the quality of both Wi-Fi and cellular connectivity enabling dual-connectivity, intelligent handover and more. Section 3.2.3 set out why this would be attractive for Wi-Fi owners.

- Break the link between user and device. Any device that the user had logged into would be able to receive calls and messages. This is already the case with, eg laptops and Wi-Fi where a user can take any laptop, log into Wi-Fi, then log into eg WhatsApp then receive the full service.
- Remove the need for a phone number. While the phone would still need its IMEI identifier, users could have friendly addresses, such as their email address, rather than an 11-digit number. This also removes issues of number assignment and number shortage.
- Make highly tailored national roaming possible since the coordinator would have visibility of multiple networks and so could move the device between then according to detailed rules.
- Remove the need for MNOs to provide voice and messaging services, which has become increasingly difficult as networks have moved toward being all IP. As with fixed networks all such services would be provided over-the-top.

Optimally connected

This section has mentioned that one function of the coordinator would be to move devices across networks so that they were always optimally connected. There are two extremes in how this might work:

1. The coordinator sends rules to the device, which then moves itself across networks in accordance with those rules and its local environment.
2. The coordinator sends network handover commands to the device, actively moving it across networks.

In practice, some middle ground is likely to be appropriate. The device will likely need to move itself onto Wi-Fi networks since the coordinator will have little visibility of what Wi-Fi networks the device can see. However, the coordinator can send the device a list of nearby Wi-Fi networks that are available. The device may need to move itself onto satellite networks since it may not have any connectivity at the point it needs to connect to a satellite. But the coordinator may move it off the satellite if it knows there are other network options. The coordinator might need to intervene more to move devices across cellular networks (national roaming) but this will depend partly on whether the national roaming algorithms are modified as part of the 6G standard.

It is unlikely that the coordinator will be highly active in managing devices, this would require continuously updated information from devices and networks on signal levels and available capacities. More likely is the coordinating network sends devices sets of rules and data on nearby networks to allow the device to make better informed decisions itself.

APIs

Cellular networks provide a range of application programme interfaces (APIs) that can deliver information useful to applications such as the current device location, device identity, one-time passwords and can request higher quality treatment. These tend to only work across specific cellular networks. The coordinator can provide "global APIs" that work regardless of the network that the device is connected to. It can do so by "translating" between different networks and even moving devices across networks when requested by an application. Applications then run at a multi-network level rather than being restricted to particular networks or technologies.

How a coordinator function might be achieved technically, politically and economically is the subject of the next chapter.

6.6 Protocols (by David Lake[54])

Current protocols make it difficult to separate the application from the transport, and as a result to easily implement the multi network coordinator concept described earlier. This section discusses how that could change.

---

[54] David Lake is an Engineering Technologist with over 25 years of network design and deployment experience, ranging from X.25 and SNA, through the era of multiprotocol routing to IP. He has extensive experience in transporting rich-media technologies across complex enterprise and service provider networks, has served on the ETSI NFV, MEC and NGP groups, IETF DMM and IRTF ICNRG. He holds a part-time researcher post in the 5G/6G Innovation Centre at the University of Surrey with a number of publications and patents.

Not much has changed in terms of network protocol support since GPRS appeared as a commercial service in 2000. The base protocol, IP, is carried in a tunnel over GTP (GPRS Tunnelling Protocol) anchored within the fixed part of the core network. New GTP tunnels are established as a mobile moves across base stations and then switched by the MNO in order to make the mobile device appear static; once allocated within an MNO, the IP address will rarely, if ever, change.

Whilst the speeds of the IP session have increased with new RF technology, there has been little evolution of the services beyond the voice and simple messaging MNOs have long delivered. VoLTE/NR capitalises on the ability to use the IP layer for SIP-based communications but remains tied to the original operator-voice service. These applications share techniques with Internet VoIP systems but remain a separate service.

There is a difference of approach from the two main network-related standards development organisations (SDOs) that explains the situation we are in and why so little has changed over the last 30 years. IETF came from a wired-Internet background where interoperability of individual components was the goal. They have never concerned themselves with the access technology, leaving that to other groups such as IEEE to provide standardisation and architectures. Protocol definitions in the IETF remain recommendations, not diktats and the group prides itself in the separation of transport and application, the so-called 'narrow waist' of IP. Neither the regulatory framework nor the economics have overly worried the IETF whose proponents have typically been the new Internet companies and services.

On the other hand, the 3GPP represents the operators and manufacturers in the highly regulated public network space, often the descendants of the monopoly national telcos. With their focus on the cost of operating the network, the need to provide services within tight national and international boundaries and a highly competitive consumer market the scope for the kind of radical change which the IETF often proposes is limited.

Because of the nature of the current IP protocol suite, changing IP addresses can be catastrophic for higher-layer protocols which derive their state from the IP network address. The whole idea of a mobile network is anathema to IP where addresses were designed to be geographical in nature. This has been recognised as a major failing in the protocol over the years with many retrofit solutions being proposed including Mobile IP or Recursive Internetworking Architecture (RINA) and some radical new ideas such as Named Data Networking (NDN) being proposed. To-date, none have found traction and the only way that the static requirement of IP can be married with the mobility requirement of 3GPP is through GTP. In fact, it is interesting to note that the SDO responsible for the design and development of IP, the IETF, had no part in the architectural or even protocol elements of the mobile network, GTP being an ETSI/3GPP invention. And yet, today, much of the Internet connectivity is via a mobile device, connected either over cellular or Wi-Fi.

Abstraction of the application from the networking stack such that the application no longer needs to see a consistent IP address and connection, has enabled applications to work around this limitation; one only needs to see how a video will continue to stream from a distant server when moving from a home Wi-Fi to a public cellular connection. The application does not care which service it is connected to other than there being an Internet connection available. The needs of the application dictate whether this is feasible; for example, a video playback solution using an HTTP-derived protocol such as Dynamic Adaptive Streaming over HTTP (DASH) will cache and pre-fetch large portions of video and the switch over from one access mode to the next will be masked due to playback buffering. It would be more difficult to build a latency and loss sensitive application to be as robust and this accounts for the oft-seen breakups in OTT applications such as Zoom or Teams.

This abstraction does not extend to VoLTE/NR and VoWiFi. In this case, whilst techniques exist to provide consistence when moving from LTE or 5GNR to Wi-Fi, this is a carrier-controlled service that tunnels the VoLTE/NR service over the Wi-Fi IP connection. There is tight coupling of the application to the transport and the switch between access methods or degradation to circuit-switched 2G voice in areas of marginal 4G/5G coverage is completely under the control of the MNO.

Originally, the IP stack was designed to have an 'Internetwork' layer that would be agnostic of the underlying IP addressing scheme, thereby enabling the lower-layer elements to change without impacting the connection. This layer was replaced by connection-orientated services such as TCP and HTTP. TCP itself suffers from both long set-up times and a sliding data window which can limit throughput. It has no concept of a 'session' that transcends the network layer meaning that every network attachment, even if coming from a pre-existing service, appears new to the two end IP stacks.

The current tunnelling approach in mobile networks is clumsy and wasteful. Considering the overall network stacks involved, the GTP which runs between every cellular base-station and the aggregation point and between switching elements in the operator's packet core is itself a UDP packet. This packet, with a GTP identifier encapsulates the entire mobile data packet, a total of two sets of protocol plus an additional header. For small data frames the ratio of header to data is highly inefficient. Furthermore, the GTP information only has relevance within one operator network – it is purely a tool for moving data between elements, not the abstraction of application and transport we need.

There are two areas where protocol development and a change in some business practices could help as we try to realise ubiquitous connectivity.

First, in terms of the legacy voice services (that is voice/text delivered through a combination of Circuit Switched 2G and VoLTE/NR/Wi-Fi), they should be seen as applications and delivered in the same manner as any other service a consumer would buy from an application store – ie MNOs should stop delivering native voice and messaging. As one looks ahead to 6G, there should be no reason to retain any 2G connectivity and therefore no reason for operators to maintain complex and expensive IMS systems or GTP switching. In the same way that the PSTN has moved to a SIP-based solution, so should the mobile networks. Secondly, promoting a base-level of interoperability between applications using open, documented protocols should be the goal; allowing the user to run those applications on whatever device and with whatever connectivity they choose. For example, it should be possible to be able to call and text between any of the current platforms such as WhatsApp, a broadband phone, the dialler on my mobile or any other web-based application. By using existing protocols such as SIP and DNS Enumeration (ENUM) records, we can abstract the application – E.164 numbers – from the service – connectivity. The MNOs become, like the ISPs, providers of connectivity.

Abstraction of application and network connection for data services requires that the problem of a lack of an "Internetwork" layer be addressed. One protocol which seeks to address this challenge is Quick UDP Internet Connections (QUIC). By providing a Connection ID above the network and session layers of the current IP stack, applications are able to associate different IP streams across multiple connections thereby enabling application mobility without having to form a new relationship. QUIC also seeks to address the slow connection set-up times of TCP, running as it does over UDP with a much-reduced handshake. Media-over-QUIC enables real-time streaming and two-way audio sessions in highly mobile environments and applications are being rapidly built that capitalise on the very robust nature of the protocol.

If these two elements – separation of the service from the transport and use of protocols like QUIC - were widely adopted, one can see how the legacy of GTP and IMS could be removed, leading to simpler, cheaper access networks focused purely on providing good quality connectivity.

6.7 Who owns the multi network coordinator?

There are some advantages in the multi network coordinator being a single global entity. This removes all issues with roaming and the need for a subscriber identity to have a national indicator (as current phone numbers do in the international prefix digits). Equally, there are many downsides including the risks of failure, national security and sovereignty, and the challenge of managing local handover decisions from a global centre. On balance, national servers are probably a better decision, with each national server accessing a global address book when needed or maintaining a copy of the global address look-up which it periodically updates from a global address server – quite possibly the existing DNS servers.

There may be exceptions – for example there could be a multi network coordinator for the whole of Europe. This would fit with EU visions of consolidated telecommunications provisions across the continent without the need for difficult merger of national fixed and mobile operators.

There could be multiple regional multi network coordinators in a country, rather than national multi network coordinators, and there is no reason in principle why this would not work. But given that the structure of telecoms networks today is generally national then keeping the multi network coordinator at a national level makes most sense.

There could be multiple multi network coordinators in a country, with subscribers choosing who their multi network coordinator provider is (as they choose who their ISP is). Alternatively, each network operator could also implement a multi network coordinator and provide this as part of the cellular or fixed access service. Multi network coordinators could compete on service level, the number of networks they can access and, of course, price. There seems little penalty in having multiple coordinators, other than the slight increase in complexity, the challenge each coordinator has with negotiating access arrangements with each access network, and duplication of effort involved.

Multi network coordinators could be implemented in software and run on cloud servers. As a result, multiple instances could easily be spun up as needed, and the cost of owning and deploying a multi network coordinator could be very low.

Messaging capabilities could be directly provided by the multi network coordinator. But it would also be possible for each subscriber to nominate their preferred OTT platform (which could change as often as they wish). A call from one subscriber using OTT provider A to another using OTT provider B would then be possible as long as the OTT providers delivered the voice or messaging in a format that all OTT providers recognised, or the coordinator translated between them.

Hence, in the most complex structure a subscriber would:

1. Sign up to a home broadband service.
2. Sign up to a preferred MNO (delivering bit-pipe connectivity).

3. Potentially sign up independently to a Wi-Fi aggregator although this may be included in the multi network coordinator network offering.
4. Potentially sign up to a satellite provider, although again this may be included in the multi network coordinator network offering.
5. Select a multi network coordinator provider.
6. Select an OTT provider.

While this may sound laborious, in practice, subscribers typically already sign up to (1) home broadband, (2) MNO and (6) OTT provider. The services (3) Wi-Fi and (4) satellite may be included in the coordinator offering. Hence, the only additional step is (5) selecting the multi network coordinator, and if there is only one per country, or if every access network operator also operates a multi network coordinator, then even this may not be needed, with the multi network coordinator subscription, for example, bundled within the MNO costs.

So who might provide the multi network coordinator? It could be almost anyone, and there could be multiple coordinators per country. Obvious candidates are the fixed and mobile network operators who could extend their role above their own networks, although this might hamper effective national network roaming which would be onto networks of their competitors. It could be a coalition of all the network operators, as, for example, the SRN is owned by all of the MNOs, which would have the advantage of facilitating national roaming while keeping each MNO as the prime owner of the customer. Other candidates are entities like MVNOs. They already are companies with billing relationships with subscribers and access relationships with operators, albeit typically just one MNO. Some "thick" MVNOs even have their own core network already. Alternatively, it could be global players like Apple and Google that already provide messaging services and identities (and in the case of Apple, satellite connectivity).

There will be plenty of interest in being the multi network coordinator provider. Allowing competition, with multiple national coordinators may be the best way to allow innovation and the best multi network coordinator operators to emerge.

6.8 *Economics*

It is critically important in any 6G world that all players in the value chain are suitably profitable. As noted in Section 2.5 MNOs are currently struggling financially due in part to an inappropriate 5G standard.

Much of the current industry financial structure is unchanged under this approach:

- Home and office broadband is completely unchanged, with monthly subscription paid to the broadband provider.
- Mobile service would be mostly unchanged, with monthly subscription as currently paid. The MNO would not need to provide voice and messaging.
- Wi-Fi would be unchanged – free in most cases, delivered as part of an overall service package (eg as part of the cost of a hotel room). There may be Wi-Fi aggregators who would require payment which could be a monthly subscription but could be bundled within a larger connectivity package or be part of the home broadband or cellular package.
- Satellite is unclear. At present the only available service is free – bundled as part of the purchase price of an iPhone. In future there may be a need to pay a subscription to a satellite provider for those who want the service, or it may be bundled as part of a larger connectivity package.
- OTT providers are generally free to users, generating revenue from advertising or the value of the user data. There is no reason for this to change.
- The multi network coordinator is a new entity with multiple funding options as discussed further below.

Hence, for most players, the economics are broadly unchanged, other than they only need provide bit-pipe services and may not need to invest as much in network enhancements, both for coverage and capacity, as they currently do.

The multi network coordinator could be financed in multiple ways:

1. Via a separate subscription paid by users in addition to their fixed and mobile subscription. While users will not want to pay more, the cost should be very low, given the very limited investment needed.
2. Via selling a complete package. The multi network coordinator could become a "one stop shop" where for a single fee a user is provided with a bundled service comprising broadband, mobile, Wi-Fi and satellite. This is similar to the bundled packages sold by some integrated fixed and mobile operators. The multi network coordinator provider then buys wholesale service in the same manner as MVNOs currently do.
3. Through being bundled with another service. For example, users generally buy home broadband from an ISP that delivers the bit pipe (often leased from a wholesale provider) and the onward connectivity, billing and customer care. A broadband provider or mobile network operator could offer a combined broadband and coordinator network service, allowing the user to add in their preferred mobile supplier, etc.
4. Through advertising or other non-payment activities in the same manner that WhatsApp is financed.

Some of these could emerge in parallel giving users a choice as to whether they prefer the simplicity of a completely bundled approach, or the control of selecting each element of their connectivity themselves.

This appears entirely workable. Current MNO cost issues are not directly addressed through the multi network coordinator concept, but as discussed elsewhere if 6G can reduce network costs in other ways, such as through increased automation, then this allows greater profitability.

6.9 *Security*

Several researchers suggest that 6G will need enhanced security compared to 5G. Broadly, if there is a security issue with 5G then intra-generational updates can be used to address it. However, if 6G is sufficiently different from 5G that new security concerns arise then these must be addressed within the 6G standard. Security, while critically important, tends to have a limited impact on network cost and hence there is no "choice" that needs to be made here – appropriate security should be added as needed.

The key difference proposed here between 5G and 6G is the true hetnet architecture, with a multi network coordinator and authentication taking place above individual networks. This will undoubtedly raise additional security concerns such as the ability to trust authentication that has been performed elsewhere. Such an architecture might also make greater use of cloud computing and storage which could add new security weaknesses. Mechanisms that address these and other security concerns that might emerge must be part of the 6G standards.

6.10 *Lowering energy usage*

One of the calls from the operators has been for lower energy usage, in absolutely terms. As noted earlier, 5G significantly increased the absolute energy use, which both adds opex to operator costs and is not environmentally friendly.

Reducing energy is not easy, especially as ever-more complexity is added to the network. The broad approaches that would deliver lower energy usage are:

1. Rely more on indoor coverage for indoor devices. This hugely reduces the power needed since the range is now short and the building penetration loss no longer applies. With something like 80% of all data transmission from indoor devices this could make a large difference.
2. Share networks. A single network, even if it has four times as many subscribers, uses much less power than four separate networks.
3. Reduce reliance on MIMO antennas. These are inefficient as they have many RF elements. MIMO is predominantly used on 5G networks running at 3.5GHz, so reducing the use of this band, taking advantage of indoor offload from (1) above, can make a significant difference.

Happily, the overall 6G strategy laid out here, with its increased reliance on Wi-Fi for indoor coverage, and on shared rural networks, already aligns well with these key options for energy reduction. None of the approaches set out by manufacturers have any coherent plans for energy reduction, and like 5G will probably raise consumption.

6.11 Legal requirements on operators

Governments and regulators place a variety of legal requirements on network operators, such as lawful intercept, preventing access to certain content, data protection, and delivering unencrypted data to security services where possible.

There are already many anomalies in this system as networks have increasingly become bit pipes and communications services are delivered over the top. Often these requirements are not levied on OTT providers such as WhatsApp, while the encrypted nature of OTT communications means network operators cannot intercept and unencrypt communications. There is already a pressing need for review and change to the legal requirements and where they are placed.

A multi network coordinator would, logically, be one place where these requirements could be levied. The multi network coordinator would have better visibility of the connectivity of a given identity than any individual network. The multi network coordinator could cascade requirements as needed, for example, requesting location information for the network where the identity was current active. The multi network coordinator could also indicate which OTT messaging platform was being used. Governments could decide whether they also wished to place requirements on these OTT providers.

Mobile networks also have a duty to carry emergency calls even if the subscriber is not normally allowed to access the network (effectively a form of national roaming). The multi network coordinator concept enables subscribers to be always connected, and so the "must carry" of emergency calls is less relevant. But must carry over mobile networks can remain in place until it is clear it is no longer needed.

Hence, a multi network coordinator would be easier to introduce if there were material review of the manner in which legal requirements were levied. But such a review is long overdue in any case.

6.12 The optimal outcome

It is clear that we need to avoid the 5G-on-steroids vision. 5G has not been a success and doubling down on it is effectively throwing good money after bad.

A better 6G vision

That vision will lead to even higher frequencies, even less coverage, an increasing digital divide and further losses for operators. All to deliver a cyber physical continuum that nobody seems to want. Or it may not even be implemented, as operators decide not to deploy, and the standard sits on the shelf.

It is arguable that we do not need 6G at all. 5G delivers more than enough speed and capacity for all that we can foresee. Intra-generational updates to 5G can provide enhancements as they are needed, including aspects such as easier implementation of AI. Operators clearly do not want 6G. But the generational process has huge momentum and stopping it now would be near-impossible. It is likely that something will emerge that is termed 6G.

If there is to be a 6G it should address the problems that concern users. These are primarily of quality coverage (including sufficient capacity, data rate and reliability to do all that they want) and lower costs. The way to achieve the apparently incompatible goals of enhancing quality coverage while reducing costs are using multiple networks and new coverage approaches including satellite, HAPs, Wi-Fi and selective national roaming. Adding network simplification and automation would further reduce costs. The key approach to achieving this is a multi network coordinator that sits above the multiple heterogenous and homogeneous networks that now need to work together.

This should be the 6G vision – always connected quality coverage at lower cost delivered through a multi network coordinator architecture and with a highly focused approach to lowering network opex.

The next chapter looks at how to get there.

7 How to get to the optimal outcome

7.1 Where we are heading

We are currently heading towards 5G-on-steroids. It is the manufacturers that have the upper hand in determining what a new generation looks like. They dominate in 3GPP and have the resources to provide the input papers that become the standards. They are well versed in publicity and knowing how to whip up hype around their ideas, co-opting politicians and influencers. They have every incentive to make 6G as expensive as they believe the market can bear in order to maximise their equipment sales. Some are likely pressured by their governments to add new features for which they can gain IPR and therefore have a global advantage, and the most fertile ground for IPR is the newest and most outlandish concepts. And for those working on the standards, delivering "better, faster" is all that they have known and hence what they will naturally default towards.

There is some evidence that the voice of the operators, arguing against 6G, has softened the position of Nokia, but it is not reflected in any way in the position of Ericsson, Samsung and Huawei, at least not in their public materials. And even Nokia's position is one of 5G-on-steroids, just not on "day one".

We are heading then for a repeat of 5G. Led by manufacturers with academics enthusiastically supporting them[55] we will be told that 6G will, really, be the generation like no other. We will hear that the 5G vision was a good one, indeed perhaps not even futuristic enough, but that 5G itself was not quite up to the task. 6G will solve that, and finally enable us all to spend our lives in the cyber-physical continuum, if only we could work out what that was. There will be no remorse for all the unnecessary spending on 5G and the failure to deliver on a vision that was going to be more transformative than electricity. Operators will continue to say that they want 6G to be software only, and, as I heard at a

---

[55] To be clear, I do not blame the academics in any way for doing this. They should research whatever they want, and if there are grants available for particular research projects then they will be strongly incentivised to take them. It is not the place of the academics to decide what the overall direction of the industry should be.

conference, the manufacturers will say, in public, "what they really mean is that once they have implemented the 6G hardware then all subsequent upgrades will be software only"!

But it will be harder this time. Politicians will be somewhat more wary of telling us that our country must win the race to 6G or face global irrelevance. Consumers will reflect on how 5G has not changed their mobile experience and be less bothered about whether their operator is quick to implement 6G. Indeed, they might we be more inclined towards the operator that offers "5G at lower price" rather than the one that offers "pay more for a cyber physical experience".

Because they are less concerned that subscribers will opt for the network claiming the "most reliable 6G" or similar, then many operators will choose not to implement 6G, or to only deploy small-scale systems say in capital cities and then wait to see if the need emerges. Perhaps in countries like China, and perhaps South Korea, government pressure will lead to significant implementation. Other countries inclined towards techno-optimism might follow suit. Hence, there will be a patchy, relatively small-scale implementation. Very few individuals will experience "true" 6G[56], either because it is not deployed in their country, or because the coverage is so poor that, unless they visit major cities, they will not be on 6G networks.

Eventually manufacturers will declare success because most will have 6G-compatible phones (just as with happened with 5G, all phones will add 6G capabilities, albeit not necessarily the new frequency bands). But 6G will have delivered nothing of value to the world. Most importantly, it will not have improved coverage or reduced cost. Of course, use of satellites, HAPs and Wi-Fi can continue and grow within 5G so the user experience may improve anyway, but not because of 6G, and the improvement will not be as great as it might have been.

Is there anything that can be done to change the direction of travel?

---

[56] It may be that, as with 5G, there is a low frequency deployment of something called 6G, which is no different from 5G or even 4G, but allows operators to claim that they have a 6G network.

7.2 *Operators versus manufacturers*

As observed, there is a huge gulf between operators and manufacturers over 6G. Broadly, manufacturers want to sell kit, but operators want to avoid buying it. Operators know from their 5G experience that they can be pressured into buying the next generation equipment if politicians and the public are persuaded that it is important.

Operators struggle to work collectively. They have an industry body – the GSMA – but operators are rarely of one voice, and the politics in countries like China means operators there are more inclined to show a positive view of 6G. Collectively, operators cannot kick the habit of wanting more spectrum, or at least options on more spectrum, and hence the GSMA tends to be favourable to 6G visions that require more spectrum to encourage regulators to make spectrum available. In fact, GSMA seems to have said little about 6G to date, perhaps a symptom of the difficulty of achieving consensus.

As noted in Chapter 4, instead of using GSMA, operators have been expressing their views through NGMN. These views are clear, but not forcefully expressed. Manufacturers could read them as weak support for 6G rather than no support at all for the 5G-on-steroids vision. Perhaps the operators are more forceful in private meetings.

Since operators will be unable to speak collectively with one single global voice, they need to work locally with willing partners. Operators need to become the proponents for the 6G that they want. There is no point them advocating that there not be a 6G, or even that it be "software only", since 6G will inevitably happen. They need to set out a vision that works for them and sell this vision to politicians, influencers such as journalists, shareholders and the broader public. By doing so, they will make it much harder for the manufacturers to gain the upper hand.

And the operator vision has the advantage of being so much more appealing. Who would not prefer being always connected at lower cost to deep immersion in the cyber-physical?

They also have the advantage of a much stronger local voice. Politicians will be much more willing to listen to their national operators, companies they see as being part of their country, rather than some large equipment provider from Sweden or Finland. Operators advertise far more than manufacturers and so have brands and channels that the public are familiar with. If operators are proactive then they are certain to succeed.

Operators should aim to work collectively in their countries, selling a clear message to local politicians and journalists. Multi-national operators might try to achieve the same across all the countries where they operate.

Most operators do not currently have the skills and experts to do this. Writing persuasive vision papers and communicating them clearly, refuting counter arguments and winning debates, requires rare skills. Operators have tended towards utilities, and some have lost the ability to strategize far into the future. There is not the time to grow such experts and hence they need to be hired. Suitable talent can be found among consultancies, with a few semi-retired experts and similar[57]. Buying in the ability to ensure that 6G meets operator needs would be a very easy investment case to make.

If it becomes clear to manufacturers that there is no way that they will be able to get operators to buy the 6G equipment that the standard is specifying, then they will change. After all, they want to sell equipment, not ensure we are all cyber-natives.

7.3 *Issues of national sovereignty*

National sovereignty – having local manufacturers or other important companies in the supply chain – is being seen as increasingly important. It is partly a reaction to the concerns over the use of Chinese equipment and partly a reflection of the global trend towards self-sufficiency. Unfortunately, it is broadly unhelpful and unrealistic.

---

[57] For which I declare a self-interest!

As an example, the Centre for a New American Security said[58]:

> The country that leads the development and deployment of 6G—the next generation of wireless telecommunications—will control critical infrastructure that is integral to global economic competitiveness, national security, and the functioning of society. 6G is thus a key battleground of U.S.-China technology competition. Though 6G is not expected to be commercially available until 2030, China's dominance of fifth-generation networks highlights the importance of planning and early market entry. The United States must act now to establish 6G leadership and ensure the development of secure and resilient networks.

There are many other similar examples. The NTIA asked[59]:

> How should the U.S. Government cooperate with like-minded countries on enabling 6G success globally? Are there existing international initiatives on 6G that the U.S. Government should consider? Are U.S. companies and those of likeminded countries positioned to be global leaders in 6G development, standardization, adoption, and deployment? What other countries or regions represent the strongest challenges to U.S. leadership in 6G? What can the U.S. Government do to enable success of U.S. companies in the global 6G market?

And the EU wrote[60]:

> The global race to 6G has begun and the stakes are high, as 5G and 6G-enabled activity are estimated to generate €3 trillion in growth by 2030 worldwide. Businesses and countries are competing to build the next level of 6G mobile networks. The competition for 6G is also motivated

---

[58] https://www.cnas.org/publications/commentary/u-s-china-competition-and-the-race-to-6g

[59] https://www.ntia.gov/federal-register-notice/2024/advancement-6g-telecommunications-technology

[60] https://www.europarl.europa.eu/RegData/etudes/BRIE/2024/757633/EPRS_BRI(2024)757633_EN.pdf

by the need to ensure leadership in the technology and ensure the EU's digital sovereignty.

Early 5G frontrunners in Asia, such as South Korea, China and Japan, have started to define their vision on 6G. India and Brazil have announced investments in 6G research and development projects (R&D). In the United States, the private sector is mainly leading the 6G debate through an industry initiative aimed at advancing North American mobile technology.

The EU is also providing support for 6G research and innovation to ensure it gains a leadership position in the sector. The European Smart Networks and Services Joint Undertaking (SNS JU) was established in 2021. The SNS JU is jointly funded by the EU budget and industry. As far as industry is concerned, the 6G Smart Networks and Services Industry Association (6G-IA) works in partnership with the European Commission on the SNS JU projects. It represents the voice of European industry and research actors on 6G, bringing together operators, manufacturers, academics, small and medium-sized enterprises and ICT associations.

These desires for "national leadership" lead to funding for research and pressure on companies and operators to ramp up involvement.

While the EU is currently supporting the 5G-on-steroids vision, there may be interest in the multi network coordinator concept. The EU white paper on telecommunications - "How to master Europe's digital infrastructure needs?"[61] discusses why, in the EU's view, deployment of 5G and fibre has not reached the goals set by the EU. It concludes that one of the reasons is the lack of a single market for telecoms, with national operators in each country. In Section 3.2.2 it sets out one of its proposed solutions:

> At the same time, the recent technological changes create an opportunity for alignment of the operations of electronic

---

[61] https://digital-strategy.ec.europa.eu/en/consultations/consultation-white-paper-how-master-europes-digital-infrastructure-needs

communications and cloud services with the development of pan-European core network operators. For example, the cloudification of 5G networks can provide significant benefits to the electronic communications network providers and allow them to leverage the same economies of scale of cloud providers by, inter alia, unifying the core network functionality of several national electronic communications networks in the cloud.

It is very unlikely that national operators would "unify" their 5G core network functionality. However, a unified multi network coordinator for the whole of Europe is much more plausible. It does not require integration of existing cores and there may be sensible business and Community reasons for promoting it. Hence, the 6G vision set out in this book might be a way to realistically deliver against some of the EC objectives. To achieve this, the EU will need to rethink their views on 6G, something that this book shows is strongly needed in any case.

The desire for national sovereignty is a fact of life and something that needs to be managed and ideally used for good. The message to those concerned about sovereignty should be:

1. There is no point in having leadership in something that is not deployed or is unsuccessful. Indeed, this is wasteful.
2. The vision set out by the incumbent suppliers is one that favours their own interests and makes it harder to grow national champions (outside of Sweden, Finland, China and South Korea).
3. The way to achieve sovereignty is to champion a different vision that delivers real benefit and changes the overall architecture of the cellular network, facilitating new entrants (in the manner that O-RAN somewhat successfully did).
4. The multi-network vision plays to the strengths of countries with satellite, Wi-Fi, HAPs, IP network and other similar skills, breaking open the current supply chain eco-system.

If this message can then help direct funding, government interventions and lobbyists towards delivering the optimal outcome then national sovereignty can be turned into a force for the good.

7.4 Challenges in delivering a multi-network coordinator

The multi network coordinator concept is one that operators of access networks might initially find concerning. Their first reaction may be that they will be relegated to bit-pipes and may lose their customer relationship. They may be concerned about "outsourcing" authentication and other critical subscriber management tasks. And most large companies are naturally fearful of change.

But there is little need for concern. The access network operators might also own and operate the multi network coordinator, extending their reach. Even if the multi network coordinator is owned and operated by a third party, users will still need a subscription to a mobile and fixed operator (in the same way that they can only access Wi-Fi if allowed to by the owner). The relationship may look more like that with WhatsApp – users pay the operators, and WhatsApp delivers the coordinating solution. Operators maintain the same customer relationship and the same billing arrangements.

If the operators own the multi network coordinator or have a tight relationship with the multi network coordinator provider, then a collective ownership among all operators might be better than one from a single operator in order to facilitate national roaming. In any case, operators are very well placed to influence the multi network coordinator specification, implementation and operation. They can control the change.

Another challenge is where the standards for the multi network coordinator get developed. 3GPP works on cellular standards and the IEEE on Wi-Fi, although they are increasingly finding ways to work better together. The multi network coordinator needs to sit above both, so it is not appropriate for one or the other to take the lead, and indeed if they did a balanced solution is unlikely. Creating a new standards body from scratch is time-consuming and likely to be enveloped in politics. The solution may be akin to how 3GPP was formed which was through the agreement of multiple national and regional standards bodies like ETSI. A 3GPP-IEEE Partnership Project could set up a jointly owned entity able to work with both partners to deliver a multi network coordinator specification. Since these bodies work increasingly closely this may not be problematic, and perhaps may not need material change.

Finally, those who are reluctant should ask themselves whether such an arrangement will happen anyway – a sort of WhatsApp take-over – without their involvement, input and leadership. Better to be the one who invents the future.

7.5 Government support is likely to be needed

Although this book has set out ways to enable users to be always connected at low cost, it may still not be in the economic interests of the operators to do this. As discussed in Section 3.2.1 unless governments (or similar) introduce a way for consumers to accurately compare operator coverage then market forces will incline operators towards reducing capex. In "Emperor Ofcom's New Clothes" my co-author and I set out at length how governments could support and incentivise operators to deliver better coverage.

A key part of this is funding. Governments already fund coverage, either directly through schemes like the UK's Shared Rural Network, or indirectly through allowing free licence renewals (and therefore forgoing auction revenues) in return for coverage commitments. More of this is likely to be needed in the future. Governments can allow operators to retain licence fees if used to meet set objectives, can auction spectrum and other rights against promises of coverage or can directly provide funding.

Governments can also facilitate, for example, as mentioned earlier, by getting a shared Wi-Fi solution off the ground or by seed-funding start-ups with novel ideas in areas such as HAPs. Regulators can fast-track satellite D2D regulations, possible finding additional spectrum to facilitate ubiquitous coverage.

None of this changes the optimal vision for 6G nor the way to achieve it.

7.6 Time for the operators to make a stand

Over the various generations the operators have oscillated from leadership to complacency which inevitably leads to dissatisfaction.

How to get to the optimum outcome

- They were very much in the driving seat for 2G and very happy with the outcome.
- Industry took over for 3G and delivered a technology the operators found hard to make work.
- The operators took back control for 4G and used it to correct the problems with 3G.
- They then sat back for 5G and got a solution in search of a problem.

It would fit this cycle for the operators to react against the way they were pushed into 5G by taking control of the 6G agenda to deliver the network and solutions that they want – ones that increase their profitability and customer satisfaction. To do so, they will need to strengthen their team, recruiting proven visionaries with persuasive abilities.

As discussed in this chapter, collective action at a global level is unlikely to succeed but national and regional leadership has every change in doing so. Especially when the choice is being always connected versus cyber-dystopia.

This book is entitled "A 6G Manifesto". The next chapter summarises the contents of the book and then the final chapter sets out the vision developed in earlier chapters as a manifesto for change.

8 The book in overview

8.1 The long summary

The world of cellular connectivity moves forward through the advent of new generations. A new generation typically arrives every 10 years, and historically has delivered around a 10x improvement in real-world data rates and, using new spectrum and improved spectrum efficiency, has provided increased capacity. From 2G to 4G, new generations were clearly delineated by a new air interface – TDMA for 2G, CDMA for 3G and OFDM for 4G. But this change did not occur with 5G, which kept the same air interface as 4G albeit with increased flexibility and bandwidth. 5G promised much – to be a generation like no other that delivered a new world of immersive communications – but has disappointed, with no new applications, no increased revenue for operators and customers that are only connected to 5G a fraction of the time and perceive no material difference from 4G. 5G did deliver more capacity, and this has been important as data usage grew from the time of its introduction to the present day.

The cellular community, led by the manufacturers and academics is now discussing 6G. But it is far from clear that we need 6G. Data growth is slowing and will plateau long before 6G is introduced so there is no need for more capacity. As seen with 5G, there are no new applications on the horizon, and even if there were 5G is capable of handling all of them. Operators do not want 6G because they perceive it will result in expense for them for no benefit. A sensible approach would be to put 6G on hold until such time that it is needed, if ever.

However, the 6G "super tanker" has already left the port. Manufacturers, governments and standards bodies have activities in place that will result in 6G arriving around 2030 and stopping these is near impossible. The arrival of 6G is inevitable. But if that is so, then at least it should be a generation that benefits users and stakeholders. It is the operators that have the best view of what users need – operators are closest to the end user, and they understand best the economics of providing cellular service. Operators had a weak voice in the development of 5G which was driven more by manufacturers and academics.

The book in overview

Understanding what users need is not hard – surveys can ask them. Such surveys suggest that reliable, always connected services at a data rate of around 1-5Mbits/s which allows typical applications is by far the most important factor, followed by cost. Factors such as higher data rates (after the basic connectivity need is met) are of little importance. Put simply, users want to have good connectivity wherever they are and all of the time. Most do not experience this at the moment, with not-spots in urban areas, uncovered parts of the country, poor connectivity on trains and often weak or non-existent cellular coverage indoors. Where there is connectivity, it can be insufficient to enable basic browsing and applications either because of weak signal or limited capacity. The challenge, then, is delivering always connected quality services while ideally reducing the cost to the end user.

The table below summarises the ways that users can be always connected, with the solutions that provide greater coverage without requiring significant spending from the mobile operators in bold.

Challenge	Solution(s)
Rural gaps	More cells, **high altitude platforms such as aerostats and drones, satellite direct-to-device coverage**
Urban not-spots	Small cells providing in-fill, **network roaming to other networks with coverage**
In-building	**Use of in-building Wi-Fi**, dedicated in building cellular networks where demand merits the cost
Travel	**Tailored solutions for trains** based on microwave links to carriage roofs and in-train Wi-Fi, leaky feeder in tunnels
Capacity issues	**Capacity expansion with 5G**, additional spectrum or cell splitting, possible demand throttling

Table 8-1 – Summary of ways of improving coverage and capacity

The table shows how being always connected can be delivered at low cost by utilising other networks including satellite, Wi-Fi and national roaming across other cellular networks. Satellite networks that communicate directly to the handset represent a new development, while new forms of high-altitude platforms such as tethered aerostats offer new lower-cost approaches to rural coverage. Wi-Fi aggregation is now happening with new standards. National roaming across multiple cellular networks has been avoided in the past, but a new standard could remove the current issues such as the long time it takes, the impact on handset batteries and the loss of IP session information. Utilising these other networks can happen with the existing 4G/5G standards but could work better if a new 6G standard focused on optimising multi-network or heterogeneous network (hetnet) operation.

Hetnet integration is not the direction that the manufacturers and academics are currently proposing for 6G. Instead, they are arguing for "5G-on-steroids" – a solution that is even faster than 5G, with lower latency, using higher frequency bands. They argue that 5G did not deliver on the vision of a "cyber physical continuum" because it was not good enough. But 6G, being even better, will finally enable us all to exist in virtual digital-twin worlds. Such solutions will be much more expensive than 5G because they will utilise higher frequency bands that have worse propagation, requiring a lot more cells. Or, if only deployed to the same degree as 5G, then 6G will only be available to a small minority. As with 5G, this very limited availability strongly disincentivises new applications since no consumer wants to rely on an app that only works 10% of the time, or less. This form of 6G will fail, just as 5G has.

The operators are pleading for 6G to be a software only, cost-reducing update which improves coverage. This would be more in line with what the users want, but the operator voice is weak, and operators have not set out how this outcome could, and should, be achieved.

Key to delivering the hetnet solution in a way that seamlessly integrates all the different networks and overcomes existing issues, is the concept of a "multi network coordinator". This is a coordinating function that sits outside of individual networks and coordinates and provides common services across all of them. It enables services such as authentication to be accessed from any network and provides centralised control of the device regardless of the network that it is attached to. WhatsApp is an early manifestation of this, delivering services regardless of the mode of connectivity to multiple devices simultaneously (eg a phone and a laptop), keeping track of the user's mode of connectivity and using common authentication across all networks. But it does not control what network(s) devices opt to connect to. The fixed network is configured somewhat in this way, with little in the way of a core network, simply acting as a bit pipe to route traffic to peering points and destination networks. But the mobile network is the opposite, with core networks of ever-growing complexity delivering ever-more services including trying to centralise voice control even when delivered over Wi-Fi.

There are some functions that need to remain within the mobile network (although not necessarily in the core). The network needs to know a user's location so that it can route incoming messages. It may need to gather data on usage for billing. It needs to manage user mobility by handing off users between cells. It needs to provide operations and maintenance capabilities by monitoring network functionality and raising alarms. But other functions including authentication, service provision and user data, could be managed externally. Indeed, they already are for roaming users, where they are authenticated by their home network rather than the network they are connected to. Messaging, including voice and text-style messages should not be delivered by any individual network but run over the top (OTT), within or even above the multi network coordinator.

Moving device control to a multi network coordinator would:

- Make it much simpler to integrate private and neutral host networks into the broader cellular network infrastructure.
- Make integration with satellite networks simpler since the multi network coordinator would be aware of whether a device was currently registered on a satellite or terrestrial network and route traffic accordingly.
- Make Wi-Fi integration simpler since the same authentication credentials could be used for both Wi-Fi access and cellular access and there would be visibility of the quality of both Wi-Fi and cellular connectivity enabling dual-connectivity, intelligent handover and more. This can be made attractive for Wi-Fi owners.
- Break the link between user and device. Any device that the user had logged into would be able to receive calls and messages. This is already the case with, eg laptops and Wi-Fi where a user can take any laptop, log into Wi-Fi, then log into eg WhatsApp then receive the full service.

- Remove the need for a phone number. While the phone would still need its IMEI identifier, users could have friendly addresses, such as their email address, rather than an 11-digit number. This also removes issues of number assignment and number shortage.
- Make highly tailored national roaming possible since the coordinator would have visibility of multiple networks and so could move the device between then according to detailed rules.
- Remove the need for MNOs to provide voice and messaging services, which has become increasingly difficult as networks have moved toward being all IP. As with fixed networks all such services would be provided over-the-top.

Of course, there are many challenges with achieving this, mostly because MNOs will be reluctant to lose control over subscribers (although they will still have a billing relationship with those using their network). It would also need a change to the structure of the standards bodies.

So who might provide the multi network coordinator? It could be almost anyone, and there could be multiple coordinators per country. Obvious candidates are the fixed and mobile network operators who could extend their role above their own networks, although this might hamper effective network roaming which would be onto networks of their competitors. It could be a coalition of all the network operators, as, for example, the SRN is owned by all of the MNOs, which would have the advantage of facilitating national roaming while keeping each MNO as the prime owner of the customer. Other candidates are entities like MVNOs. They already are companies with billing relationships with subscribers and access relationships with operators, albeit typically just one MNO. Some "thick" MVNOs even have their own core network already. Alternatively, it could be global players like Apple and Google that already provide messaging services and identities (and in the case of Apple, satellite connectivity).

Bringing about this vision requires that operators, of all types of networks, take a strong lead in telling politicians, journalists and the public about a vision that will bring clear benefits to all users, and then pressure manufacturers to ensure that the standards deliver appropriate solutions.

Operators will likely struggle to speak collectively with one single global voice, so need to work locally with willing partners. Operators need to become the proponents for the 6G that they want. There is no point them advocating that there not be a 6G, or even that it be "software only", since 6G will inevitably happen. They need to set out a vision that works for them and sell this vision to politicians, influencers such as journalists, shareholders and the broader public. By doing so, they will make it much harder for the manufacturers to gain the upper hand.

And the operator vision has the advantage of being so much more appealing. Who would not prefer being always connected at lower cost to deep immersion in the cyber-physical?

Most operators do not currently have the skills and experts to do this. Writing persuasive vision papers and communicating them clearly, refuting counter arguments and winning debates, requires rare skills. Operators have tended towards utilities, and some have lost the ability to strategize far into the future. There is not the time to grow such experts and hence they need to be hired.

If it becomes clear to manufacturers that there is no way that they will be able to get operators to buy the 6G equipment that the standard is specifying, then they will change. After all, they want to sell equipment, not ensure we are all cyber-natives.

While the 6G super-tanker will arrive at a destination, it is not too late to change which destination it heads for.

8.2 The short summary

The world does not need 6G. Indeed, apart from the extra capacity it provides, the world did not need 5G. When well-connected we have more than enough speed and our need for data will soon plateau. Despite not needing it, we will get 6G. There is too much momentum behind the cellular generational approach with standards bodies planning for 6G, governments funding research and manufacturers driving it forwards in the hope of a revenue boost. The 6G super-tanker has left the port and will arrive at a destination. But we can still choose what destination it arrives at.

Manufacturers and academics would like 6G to be "5G on steroids" – even faster than 5G with more capacity and lower latency, so we can all live in the "cyber physical continuum". Given that 5G has disappointed then this vision is highly inappropriate. Consumers are clear on what they want – to be always connected with sufficient speed and capacity to undertake typical tasks such as social media and video calls, but at a lower cost than today. For many consumers there are still areas without coverage, not-spots in towns and cities, poor indoor coverage and terrible connectivity on trains. Operators broadly share this vision for 6G.

To deliver better coverage without raising costs requires lateral thinking. It can be achieved by using existing networks – Wi-Fi to deliver indoor coverage, national roaming across other cellular networks to resolve not-spots, and satellites to deliver deeply rural coverage. Lower cost ways of delivering rural connectivity using new forms of high-altitude platforms such as tethered balloons are emerging. 6G should be about properly integrating all these networks to deliver a seamless "hetnet" service.

The best way to deliver seamless heterogeneous network service is through a new multi network coordinator function – a platform that routes calls and messages to and from devices, optimises the networks they attach to, manages factors such as authentication, and links to OTT messaging providers. 6G should standardise this multi network coordinator and ensure the underlying networks are optimised to work well with the coordinator. The coordinator could be provided by existing operators, MVNOs, or companies like Google and Apple. There could be multiple competing coordinators per country.

Changing the 6G destination from 5G-on-steroids to true hetnets requires operators to actively advocate with manufacturers, politicians, journalists and more. But they have a much more appealing vision. Who would not prefer being always connected at lower cost to deep immersion in the cyber-physical?

While the 6G super-tanker will arrive at a destination, it is not too late to change which destination it heads for.

9 The manifesto

Mobile connectivity is critical to our way of life. Being connected everywhere, all the time, with sufficient quality to do what we want to and need to, is increasingly important to us all. And this connectivity also needs to be affordable to all. Current solutions leave much to be desired.

We want to spend our lives in the real world – connectivity should enhance this rather than be focused on enabling dystopian cyber-physical metaverses.

We are names, not numbers, we want freedom from having a mobile number, to be able to use identifiers of choice and to use multiple devices to access our digital worlds.

We have not benefitted from 5G, despite being told that it will transform our lives, and we have lost faith that the existing companies and methodologies will deliver what we want rather than what some researchers believe we need. We want our voice to drive 6G in a direction that makes the world a better place.

We believe that being always connected with good quality can be achieved by making the best use of already available resources and we want 6G to seamlessly link together multiple cellular networks, Wi-Fi and in future satellite constellations in a way that delivers coverage everywhere.

We will respect and trust operators that proactively drive 6G standards in this direction, putting our interests first, and governments that proactively change regulations and incentives to facilitate benefit to us, their citizens.

6G should be for the many, not the few.

Index

3GPP, 9, 10, 11, 12, 103, 114, 121
5GIC, 72
5G-on-steroids, 3, 49, 51, 59, 64, 84, 85, 86, 91, 93, 112, 114, 116
6G Flagship, 73
6GIA, 76
6GIC, 72, 73
aerostats, 27, 34, 125
AI native, 88, 89
airport, 60
American Bandwidth, 31
Apple, 6, 35, 36, 39, 57, 58, 108, 128
authentication, 32, 99, 100, 121, 126, 127
Bluetooth, 64
Brookings, 19
BT OpenZone, 31
Canada, 17, 39
capex, 4, 21, 44, 46, 91, 94
Carrier Aggregation, 52
CDMA, 8, 9
Centre for a New American Security, 118
Cisco, 64
CMA, 24
CPE, 53
C-RAN, 89
cyber-physical continuum, 47, 48, 49, 85, 114
D2D, 33, 35, 36, 37, 38, 40, 94

DASH, 104
data protection, 111
David Lake, 102
digital twins, 48, 50, 59, 75, 82
DNB, 44
DNS, 105
DNS server, 100
E.164, 105
EduRoam, 31
eMBB, 80
emergency calls, 112
Emperor Ofcom's New Clothes, 5
energy, 110
ENUM, 105
Ericsson, 47
ET Telecoms, 13
EU, 61, 76, 78, 118
Europe, 106
factory, 43
FWA, 34, 53
Gartner, 21
Germany, 17, 76
Google, 108, 128
GPRS, 9
GSMA, 116
GTP, 102
GWS, 23
HAPs, 27, 33, 34, 35, 40, 46, 94, 112, 115, 120
HTTP, 104
Huawei, 49

Index

IEEE, 103
IETF, 103
IMEI, 101, 128
IMT-2030, 79, 80, 82
In-building, 29
Intelligence of Everything, 50
Internet of Senses, 48, 49
IoT, 63
iPhone, 35, 57, 90, 94
IPv6, 5
ISP, 106, 109
ITU, 79
lawful intercept, 111
legal requirements, 111
lidar, 90, 91
LightReading, 71
LoRa, 64
Malaysia, 44
Marconi, 135
Meta, 56, 58
metaverse, 13, 48, 49, 56, 58, 59, 84, 85, 91
MIMO, 20, 89, 92
mMTC, 81
mmWave, 41
Mobile IP, 103
MOCN, 42
Motorola, 135
MRSS, 52
MS, 36
MSS, 36
multi network coordinator, 106, 111
multi-network coordinator, 98
must carry, 112
MVNO, 45, 108, 128

narrow waist, 103
national roaming, 3, 28, 29, 42, 46, 94, 101, 112, 121, 125, 128
NDN, 103
network slicing, 20, 21
neutral host, 42
NextG Alliance, 75
NGMN, 65
Nokia, 51
NSA, 9
NTIA, 118
NTN, 35
Ofcom, 25, 135
OFDM, 8, 9, 71, 73
Omdia, 44, 57
OpenRoaming, 31
Opensignal, 17, 23, 24
O-RAN, 89, 120
Oulo, 73
port, 43
private networks, 42
propagation, 17, 18, 87, 135
protocols, 102
Qualcomm, 13
Quantum, 68
QUIC, 105
radar, 90
Rayban, 58, 59
RINA, 103
RIS, 87, 92
rural areas, 33
SA, 9, 10, 15, 21
Samsung, 6, 11, 53
Saudi Arabia, 18
SCS, 39
SDOs, 103

security, 110
security services, 111
sensing, 82, 89
SIP, 102
SK Telecom, 14
SLAM, 90
SNS JU, 2, 76, 119
South Korea, 14, 17, 87, 115, 120
sovereignty, 78, 117, 120, 121
SpaceX, 36, 40
SRN, 28, 32, 43
Starlink, 35
Steve Jobs, 15
TDMA, 8, 9
The 5G Myth, 5, 12, 16, 91, 135
The End of Telecoms History, 5, 8, 15, 16, 41, 93

THz, 73, 74, 86, 87, 90, 91, 92
TowerCo, 45
trains, 40, 41
Transforma, 62
UK, 17, 23, 32, 34, 71, 135
unencrypted data, 111
urban areas, 27
URLLC, 80
value chain, 108
Virgin Media, 23
Vision Pro, 57, 58, 59
voice over Wi-Fi, 30
VoLTE, 102
Washington Post, 13
WhatsApp, 30, 97, 101, 111, 121, 122, 126, 127
WRC, 39

About the author

About William Webb

William is an independent consultant providing advice and support to a wide range of clients on matters related to digital technology.

He was CTO at Access Partnership, one of the founding directors of Neul, a company developing machine-to-machine technologies and networks, and CEO of the Weightless SIG - the standards body developing a new global M2M technology. Prior to this William was a Director at Ofcom where he managed a team providing technical advice and performing research across all areas of Ofcom's regulatory remit. He also led some of the major reviews conducted by Ofcom including the Spectrum Framework Review, the development of Spectrum Usage Rights and cognitive or white space policy. Previously, William worked for a range of communications consultancies in the UK in the fields of hardware design, computer simulation, propagation modelling, spectrum management and strategy development. William also spent three years providing strategic management across Motorola's entire communications portfolio, based in Chicago. He was President of the IET – Europe's largest Professional Engineering body during 14/15.

William has published 19 books including "The End of Telecoms History", "The 5G Myth", "Our Digital Future" and "Spectrum Management"; over 100 papers, and 18 patents. He is a Fellow of the Royal Academy of Engineering, the IEEE and the IET, a Visiting Professor at Southampton University, a Board Member of the Marconi Society and a non-executive director at Motability. He has been awarded three honorary doctorates and the IET's Mountbatten medal, one of its highest honours, in recognition of his contribution to technology entrepreneurship.

Made in the USA
Las Vegas, NV
13 October 2024